Master in a Miniskirt

Practical Workbook for Spiritual Living

by

Ria Panen Godesberg

Dedication

This book is for all seekers who genuinely endeavour to become the real beings they are with the objective of making this world into a better place for All. I feel fortunate to have some wonderful examples in my close circle: my daughter Erinda, my soul sisters Amina, Carolina, Cynthia and Sharon. May your intentions be always pure and the outcome a snowball effect.

ISBN: 978-1-291-92517-3

© 2012 Ria Panen Godesberg

All rights reserved.

No part of this book may be reproduced, stored in a retrieval system, or transmitted by any means, electronic, mechanical, photocopying, recording, or otherwise, without written permission from the author.

MASTER IN A MINISKIRT / THE CHALLENGE / THESE TIMES WE LIVE IN

Contents:	Page
Introduction	7
Prologue	9
Chapter 1. The Challenge of Being Oneself	11
Chapter 2. Daily Life as a Way to Inner Growth	15
1. The Butcher, the Baker and the Candlestick Maker	15
2. Giving Compliments	17
3. How the Mood Affects	18
Chapter 3. Being a Red Rose	23
1. What about Envy?	23
2. Giving	26
3. Karmic Chain	29
4. Red Roses and Grey Roses	30
Chapter 4. Small Things in a Big Context	33
1. What is Your Purpose	33
2. Being Spiritual	35
3. Energy is Only One!	37
4. Being an Example	42
Chapter 5. Rivalry	44
Chapter 6. The Dark Forces	49
1. Ahriman and Lucifer	49
2. Elemental Beings	53
3. Physical Discomfort	53
4. Desperation	54
5. Isolation	55
6. The Bond of the Dark Forces	60

Chapter 7. The Heart Chakra   63

     1. Being Who We really Are   68
     2. The Purpose in / of Your Life   68

Chapter 8. Enlightenment   71

Chapter 9. Communication   77

     1. Various Ways of Communication   77
     2. Communication of Human Beings with Human Beings   78
     3. Taking Your Time   80
     4. Being Who You Are   81
     5. Facial and Body Expression   83
     6. Communication through Thought   85

Chapter 10. Clairvoyance   87

     1. Akashic Records   87
     2. Seeing Colours   90
     3. Looking Inside the body   91
     4. Practicality on a Day to Day Basis   92

Chapter 11. Clairvoyance and Children   97

     1. Loosing Clairvoyance   97
     2. Thinking   99

Chapter 12. Meditation   107

     1. Meditation and Centering   107
     2. What is Meditation?   107
     3. Helpful Ways to Start   108
     4. Visualisation   110

Chapter 13. Horses?!   115

     1. Discrimination   118

Chapter 14. God and I — **123**

    1. Where It All Began... — 123
    2. I am It, It is I, It is We, We are One — 126
    3. Who am I? What am I?
       Is there a Difference? — 127
    4. How I see Myself and God — 128

Chapter 15. Epilogue — **131**

    1. Live as if You are Celebrating — 131

Acknowledgements — **134**

Introduction

For 6 years I have been busy writing this book. There were other people involved and then un-involved, giving me more reasons to not write this than to write it. Each person has *his or her* own idea of what I must write. So I decided to do what I feel I must do, which is to give you what I have to give. With no more talk about how to start, no more procrastination, there is no time to lose.

Giving what I have to give is what I have dedicated my life to: to give my Love to humanity. This book is a gift to you. It does not pretend to be the only guidance a person must have. It is nothing more and nothing less than my gift to you: a help to understand some things you might not yet understand and to be an inspiration to set your life on a path that will make you feel better, and will make the rest of the world feel better as a consequence. The reason I will do this quickly, now, is because I feel a tremendous urgency.

After I had written my story about the first 50 years of my life (up to 2005), I started this book. Now there is nothing left of what I started with. Many people changed, contaminated or censured my texts. I feel no bitterness, no regret. It has taught me a second time in my life that, although I love to work in a team, I have to keep the threads of my creation in my own hands; otherwise things do not work properly. Why is this? Very simply because those people do not see what I see. They cannot help it, and I cannot change it. Today is a very fresh beginning in 2011, and I start to write this new text.

I ask you to take in account that English is not my native language. At this moment in my life I find myself in the situation where, whatever language I speak, it is not my own. Even my 'own' Dutch, has become a foreign language to me. I do not find the words because I do not speak it very often. So, if you decide that this book is not good because of the incorrect use of the language, this book is not for you. Through this book I want to propagate tolerance, love, peace, goodness, brotherhood and sisterhood.

And forgiving my mistakes in the use of this language would be a wonderful beginning… As you might really want to read what I have to say… And if you have finished this book, thinking that you have understood it, or did not understand it, please do not throw it away, as you might want to read it again later; and understand things that you did not understand before, or see things from a different angle or even understand things at a far deeper level.

This book is meant to be a book of hope, the first of a series, which can help you through these 'not so easy' times. Just be prepared for surprises and let yourself be led into this adventure called life. Thank you for your trust.

Prologue

This book talks about how each person can consciously be a part of creating the present for all humanity. I want to show you how everything you do or do not do matters, because it has a reverberation in creation. It shows you how you can use simple ways of creating a better world for yourself and everybody else at the same time. It shows you that responsibility is actually freedom. It shows you that you are timeless. And it shows you how, even if we are few people, we actually can change this world dramatically. I hope I have managed to make you curious enough to be taken on the adventure that reading this book can be. Maybe you will find it interesting enough to share it with others.

Chapter 1: The Challenge of Being Oneself

Madrid 1993. It is June, most people are preparing for their holidays. The feeling of doing something spiritual, any kind of seminar or course, is not very present. It is steaming hot and people don't feel like sitting inside, and especially not at the weekend. The tarmac on the streets at midday makes the soles of the shoes stick.

The inside of most flats is muggy and gives the feeling of a prison with too many people. The noise of the surrounding flats gets accentuated as everyone has their windows open, their televisions and their radios and their cassette players on, and on top of it, trying to get heard by screaming to each other over the noise they create.

So here I am on a Friday evening, prepared to give my talk. The title is 'sexual energy'. A risky title, in a risky place, at a risky time of year. The Eco-Centre prepared the restaurant where the talk would take place, for 50 people. I was told that it would be really great if 50 people attended, especially at this time of the year. 'The talk starts at 19.30 h. Not a very good time as many people do not finish working until after 20.00-20.30 h.'

I took all this information in and told myself, 'No problem, even if there are 7 people there and they really listen to what I have to say, understand it and put it into practice I will have succeeded'. I felt good, calm and prepared, I *am* this material, I don't have to start to remember and reproduce.

At 25 minutes past seven I am in front of the door, waiting to enter the room. Yet, I cannot enter. The room is packed!

There are people all over the place, at the tables, on the tables, underneath the tables and even on the podium, where I am supposed to be, so that the only unoccupied place is where 'my' chair is. Wow, what a shock! There were almost 500 people in this room! All of a sudden I get really nervous. What do I do with so many people? I was not prepared for that... Was I?

I made my way, with difficultly and carefully, through the crowd, feeling all the eyes on me, and go onto this stage. I asked someone to take the chair off, so I could have a space to stand, and welcome the crowd. Then I told them that I had become very nervous, seeing there were so many of them.

Something happened... they all grew soft and understanding. Yet most of them did not know yet that *I* was the one giving the conference.

Something happened in me also: I knew that it did not matter whether there were 50 or 500. I had something to give that was big enough for 500 people, or even more, and it was not limited in a way that only 50 people would be able to receive this gift. I started to feel comfortable and could start giving. Of course I felt compelled to tell the audience this, and they were thrilled.

They were very expectant, as I was announced as 'Master over Life (Zen)' and as the incarnation of Loo Pingh, a female spiritual leader from 4.000 years ago from what we now call Tibet. That caused many expectations. And here I am, a woman with long blond hair wearing a miniskirt and a sleeveless top. Very modern, almost provocative, especially for this big city where I have a reputation to 'defend'. Being a female Master is a challenge in itself; being dressed as I am, and at my age, is a CHALLENGE.

I started to talk and in the beginning, the first 2-3 minutes, there was a lot of shuffling and shifting and whispering. Yet from the fourth minute they had forgotten my miniskirt and long blond hair, but felt in their hearts the truths I had to convey.

This is a challenge every time: To be one of them, of *you*. Not to stand out with 'decent' clothing as masters 'should'; not to fit the expectation that the people have of a master. Not to keep at a distance like most professionals do, whether spiritual teachers, bosses of companies, or medical practitioners. Most of them prefer to keep people at a distance, as personal contact always includes mistrust and conflict as that is part of the human nature (still).

This is the story of a very difficult path. Because not behaving in a way that people expect me to behave nor being dressed in a way that people expect me to dress, makes things harder.

It is a challenge not to conform to the accepted norms without making it a provocation, although it results in being provocative.

Yet it is wonderful to see that through love, and not dressed in conventional wrappings, through true support, people find the way to themselves, become what they really are, and find their full potential. It is a challenge I will not do without to make it easier for myself, as that would mean a corruption of the giving of true love. Because true love is new, original and unconventional, all at the same time.

Now you think: that is all very well, but why is she telling me this? I tell you this because each person, without exception, can become a Master in Love.

It is possible for each person to make a change.

It is possible for each person to play a decisive role in making the world a better place.

And that is where I would like to start.

Chapter 2: Daily Life as a Way to Inner Growth.

As you know, this book is about walking a spiritual path. So what is spiritual?? Or spirituality?

Please note first, that a person who knows lots about spiritual techniques is not necessarily a spiritual person; nor is it true that a person who knows nothing about spiritual techniques is *not* a spiritual person. Many people telling that they have a high spiritual level, often forget that little things are important, they are *the* important things.

I prefer not to define what being spiritual actually means; yet when you finish reading this chapter, I genuinely hope that you know!!

'The Butcher, the Baker and the Candle Stick Maker'

Let us go back to the title for a minute: Daily life as a way to inner growth means the very small things of life matter.

If I were to stand in front of you and give a talk, something I do very often, the very small things of life are for example that I am looking at you and I am in fact *seeing* your faces and not just sort of doing a sweep over your faces and you don't actually know whether I am seeing you or not.

I would like you to know that in fact I *am* seeing you and I know that all of you are here; I know each of your faces and afterwards I might not remember your names, because it is a bit much all at once but I know you and I respect you and I am very grateful that you are here and very grateful that you are obviously somewhere on the same line of this path of wanting to do something in this world to make it a better place.

So that is what *I* do in a conference or in a talk. And here comes your turn:

This is something so simple you can do in your daily life: presently look at the one who sits in front of you and/or next to you and just say 'Hi'.

The opposite happens so often: Everybody kissing as a habit 'Hello' and muak-muak (kissing sound) and they don't look at the person while hugging and they haven't really registered who the person is; they don't even notice that person.

That is why often I say I am not doing the kissing bit; not meaning I am not kissing in general but I am not doing the kissing as a habit.

I like to *see* the person, make contact, so if the person automatically does the kissing bit without really being present in his/her meeting with me, I will stretch out my hand. I will look in the eyes while saying: 'Hi, I am Ria' or 'nice to meet you' or whatever because it is the *meeting* that is important.

The kissing bit, the hug, the shaking hands are all just symbols; images of something and if we forgot what we want to say with them, it is just better to leave it and start *meeting* again which means that we *meet* somebody every day, really make contact with somebody!

Meet this somebody; go to the baker, look the sales lady in the eyes and say 'good morning' or 'good afternoon' or 'thank you' and pay, but not out of habit; as just a simple polite pattern that we have, but realize that you are actually grateful to this woman for giving you the bread, giving back your change and do not just take it for granted that she is there.

You will see: it makes a change. She used to feel like a machine; probably started to act like one. And we, the ones who are not really taking notice, are responsible for this. She might start to treat her kids like a machine, and her husband, her parents, her 'friends'.

Have you ever thought of that? That could happen. It can and it does! Each of our actions (talking is action) has a result, a consequence. But we don't think about these things.

Taking for granted is *not* what we must do in daily life. The same counts for all of us who go to listen to a talk, class, workshop, seminar or conference: why would we go there if we were not interested?

Then tell me, why are there so many people sitting at such an event with an air about them of: 'I know all this, it is totally uninteresting; I am not listening yet I stay until the end' (yeah yeah ☺). For the person who leads this event it is wonderful to look at people's interested faces!!! Why would we not want this person to experience this pleasure?? Why would we want to make it more difficult for her/him? Are we envious?

Show your interest; when afterwards you have questions you will not have to swallow your pride (or get stuck in your pride).

Giving Compliments

There are many situations happening in our daily life, in which we don't really realize what we are doing. For example, if somebody comes to your place (or if you go to visit at somebody's place) and the people say 'oh beautiful curtains' or 'you are looking great!' or 'oh my God, this is fantastic'.

Do you know this feeling of *'what is this about??'*

Is this about me?!

Because very many people like to constantly give compliments not because they are truly talking about you and what you have done so beautifully, but they merely want attention. They want to convince us that they are good people, aching for appreciation.

So it would be good if we learn not do this, calling attention in this sneaky way. When I see somebody and I say 'Gosh, you are looking beautiful'… why do I say this??!!

Because I want to make that person feel good because of the *person* or do I want to call attention?!?

'I want this person to like me', is that why I am giving a compliment? So is giving a complement meant as a benefit for *me*? That makes it a whole different story; it is important to recognize this, which means: we need more awareness to prevent impulsive behaviour. If we need attention, let us ask for it, if we want the other person to be aware of our appreciation for her/him, let us do it cleanly.

Spontaneous behaviour of course is wonderful, but not impulsive behaviour since impulsivity is totally subconscious. Through this behaviour without awareness we create a lot of karma.

The difference between spontaneity and impulsivity is that spontaneity comes from the heart (I will explain how living from the heart works is a later chapter) and impulsivity comes from the belly, the lower energy centres of our body, which is more like instinct.

I would like to give you a few more situations of trivial, yet real situations and the consequences.

How the Mood Effects…

Let us start this story with talking about a day after having had a night's sleep.

Let us take the example of the I-person being a woman, wife of somebody and mother of 2 children.

So I wake up and feel in a lousy mood. I stump down the stairs and start to throw some pans on the hob, plates on the table and bowls with cereal or whatever the family has for breakfast.

The kids have not shown up yet. I yell towards the stairs: 'get your behinds moving, you are late and breakfast is already on the table' all this in a really loud voice. The kids loathe this voice. They already feel rotten, and become slower coming down, as they want to spend as little time with me as possible. They feel that I am unfair; why do I have to yell at them, when there is as much time to have breakfast as any other day? They come down, don't speak to me, as they don't want to make a mistake. 'Can you not say good morning mum when you come into the kitchen?' I lash out at them. 'Good morning mum' they say with timid voices. 'Just finish your breakfast and get going!!' is my answer. They do, they eat quickly, put their jackets on and disappear as fast as they can, with a quick mumbled 'bye mum'.

I mutter to myself: 'how is it possible that they are such brats? Can they not behave better towards their mum?'

Then my husband calls me from upstairs: 'do you know where my brown shoes are?' 'Why don't you look after your shoes yourself? Why do you have to bother me with your stupid shoes?' Of course this was me, still in this foul mood. My husband comes downstairs and does not kiss me as he does not want to, and does not even remember that that is something he could do in the mood I put him in.

Maybe he is a conscious and sensitive guy and in that case he does not dare to even come near me. And I feel very offended... 'What is wrong? Can you not give me a kiss? Am I not important to you anymore?'

My husband apologizes: 'when you shouted at me like you did, I thought I'd better leave you in peace.' (kiss, kiss) I turn halfway away as I don't really like it. 'I did not shout at you!' My husband quickly has a coffee pretending not to notice and I tell him that I don't like it when he accuses me of doing things I don't do at all!!...

How does this go on?

One of my children, my son, behaves badly towards another kid in class. They fight, the teacher tells him off. My son talks back harshly. The teacher asks him: 'what the matter with you?' My son tells him that that is none of his business. The teacher feels frustrated and tells my son to leave the classroom.

My other child, who is smaller, gets bowel problems. During class she asks if she can go to the bathroom. The teacher does not allow her so she almost pees herself and just makes it, without permission, to the toilet. The teacher punishes her; she feels unjustly treated and goes to the headmaster. The headmaster calls her teacher and tells her off. The teacher feels bad because of the mistake, pissed off because she feels she has been exposed in front of her pupil and inferior towards the headmaster. Three negative feelings that make her behave accordingly.

In the mean time my husband has gone to his office. The phone rings. 'What?!' 'Sorry boss, Mr. so-and-so is calling. Can I put him through?' 'No, why do you have to disturb me? Tell him I have no time for him!' 'Sorry boss, you asked me yesterday, to call him as soon as you'd come in, so I did'. 'I have decided differently! You have a problem with that? If you do, you can look for another job!'

At midday the secretary has a date with her boyfriend. He invites her out for lunch. When they arrive at the restaurant of his choice she asks him why he always has to choose restaurants that she does not like. He is totally confused, as she normally likes places like the one he has chosen. And his good mood is gone.

My son's teacher goes home. He comes in without a word so his wife asks him what's wrong… He tells her to shut up and leave him alone. He has enough problems without her…

My daughter's teacher goes home and cries all afternoon. Her family is exasperated as there is no way she wants to talk…

Ok, I think you got it. *I* have a bad mood; I live the bad mood and pass it on to my family. My family passes it on etc. etc. Pure snowball effect.

So you think now that I am going to tell you that you are not allowed to have a bad mood??

That would be insane. You have a bad mood whether you like it or not. Bad moods are not always preventable. That is not a big deal. The big deal starts when you decide what to do with it. To say to yourself: I HAVE A CHOICE!!

Let us take the same day, with the same situation of the I-person getting out of bed with a bad mood…

Brrrr, what a day! I feel like throwing things, shouting and blaming someone else for how I feel. I'd better be careful, as no one else *is* to blame for this. I hope I will be able to keep it in.

Let's go to the mirror and give myself a nice smile: a smile that convinces me that I am a friendly person. (*smiling in front of the bathroom mirror*) Not too bad, actually it feels better already.

I walk towards my wardrobe and instead of putting on my shabby tracksuit I put on a happy coloured t-shirt and nice fitting skirt and after checking in the mirror I already feel better.

I go down the stairs and start making breakfast. Very purposely I put the bowls and the rest of the breakfast ingredients on the table very carefully. I make sure I do it with as much love as I can; as I am putting it there for those I love, even if I feel like killing someone.

I hear my kids upstairs talking and so I give a cry upstairs: 'please hurry up loves; you don't want to be late!' and they are down within minutes, giving me a hug. I tell them that I might not be so nice, as I feel rotten; they give me a compassionate look and my son even says: 'don't worry mom, I have that sometimes too. I then shout at my shoes and feel better afterwards'. 'At your shoes?' 'yeah, I just don't find them fast enough and it feels great, although they don't get any faster, as that depends on me, but the shouting at them still helps!' What a great idea this son of mine has and I notice myself smiling while thinking this.

They finish breakfast, go to school and give a quick shout upstairs: 'bye dad!' Then a hint of a kiss on my cheek, 'bye mummy', and they are gone. Happy as a butterfly.

From school I hear that day that they have been very cooperative and helped the teacher to help a couple of kids with learning difficulties. How wonderful. My husband comes down and I tell him about my bad mood and apologize in advance for any misfits of reactions I might have. He takes me in his arms and hugs me closely, telling me he loves me anyway. By now I feel a lot better already. So does everyone else and I don't think I need to tell you how it went at work of my husband...

You got it, right?

Please realise, I don't have a son and a daughter, it is not *myself* whom I took as an example. The *I*-person is fictive. But you understand what I want to make clear.

What we do with our emotions *does* matter!!! Always!!! You cannot prevent *being* afraid, jealous, angry, irritated or whatever. But you *can* decide how to react to it.

Now I don't want you to think that I don't know what it means to have, for example, a 13 year old daughter who wakes up in a bad mood almost every morning or at least that is what is seems like; because I do.

And that it might not be possible to be so nice when part of the family is also in a foul mood, I know that one too... then just ignore them and don't have arguments, as *that* will only lead to more discord. Bite your tongue, or if you can't anymore, just start crying, don't hold back, for that is the best. Being how you are is always the best. With that I mean to say that if you feel rotten, desperate etc. you can express it, yet be careful about freely letting your emotions loose, as they might become fire-spitting dragons which eventually spit back at you.

Let us have a look into a hospital. There is a loving nurse who comes into each room with a beautiful smile on her face and a friendly greeting. The days she is there the patients have less pain, less complaints. Her friendly presence will just do.

As you can see, just that little example shows how apparently small things are not small at all. Whatever we do or think, it all has a reverberation into the world and as such effects the whole. O yeah, what *you do* affects the whole world.

Often when one of my family members was angry and I felt that the reason for this had really worn off, I would ask that person to smile. There always was resistance. After I would have asked if she/he really found she/he had a right to remain miserable and spoiling the mood for everyone instead of feeling better her/himself, usually after a while they would smile. Remember this when you feel negative, whether angry, envious or whatever, and smile... for you have no right to be negative, plus... it feels so much better to be positive again.

It is very powerful what a smile does within our whole system, just lifting the corners of the mouth... does this prove there are energy points that while activated, can cause change (☺)?

That is something really wonderful. Because after this example you hopefully will become determined to be loving, friendly, helpful and so on. And that goes into the world. It will affect everybody. You will be thrilled that you are doing good.

Chapter 3: Being a Red Rose

What about Envy?

Will you allow me to give you some more examples so you can see where, in which situations, and in which ways, we throw our little bombs of dynamite?

Do you have friends or acquaintances who sometimes have said to you, while talking about some kind of teacher or therapist: she is really good, but soooo strong!!

Do you realize what has been done just here? <…but soooo strong….> What kind of underlying message is conveyed here?

This person is trying to make you believe that the teacher or therapist is not the person to go to as she/he comes across far *too* strong. Yet this person does not want to appear as someone who gossips, therefore speaks as if giving a complement.
This kind of comment usually comes from the place of envy.

Or what do you think of this one (happens a lot amongst women): 'Jesus, do you look tired!'

Listen to the difference: 'Hi there, nice to see you. Tell me, is all ok? Did you not sleep so well last night?'
What happens in the first case is that the person does not dare to tell you she/he doesn't like you, yet they want you to feel uncomfortable, so in their cowardice they *pretend* to be nice with a sting…

Yet in the second way we can 'hear' the compassion so there is love. We will not feel offended. We can take a lot when there is love…

Another one: when I was your age I did not have those wrinkles yet…

Or: interesting colours your grey and this hair dye you use; how they mingle…

Or to someone wearing high heels: 'I always wear flat shoes; they are so much better for ones back…' Then we have envy combined with the pretence to be modest, which is false modesty.

Our society is full of false modesty. Let us make an example: An enormous banquet, the room full of people and I am there too, one of a hundred people. First there is a talk and then afterwards food. Very often a lot of people in actual fact come for the food and not for the talk. As soon as the talk finishes someone announces: 'the banquet is opened'.

Nobody is going there and everybody is dying for food.
Nobody wants to be the first to take a plate, so *I* do.
What do I get from these people? Not from all of them but from many of them and usually from the same ones: 'it is always *you* being the first one to get food'.

Now this is a bit hurtful and I could feel offended yet I respond in these cases: 'No, I am the first one that breaks the ban because in actual fact all of you would like to run as what you come for is the food, and I am merely making it easy for you!'

So I go there, take a plate and because then the ban breaks and everybody rushes to do the same, I just step back and stay there and wait and am often the last one to actually serve myself food (*that* nobody notices…). But that is not the point. I am getting negative comments, reproaches.

Why?

Because I stand out, I don't mingle and disappear in the crowd.

Can you feel the underlying nastiness in these exclamations? And I tell you: all these comments come from envy.

When I arrived in Portugal, I arrived in Lisbon, where everybody dresses kind of dark: and even the streets, the buildings, are kind of insipid… I arrived in the clothes like I normally wear, full of colours. This caused most people to look at me.
Interestingly in this country the people don't really look with this envious look. They just came to me and looked me up and down like children do, and then, when they had seen it all, went again; some smiled, some did not, yet no nasty looks at all, just curious.

Yet when I visit the U.K. or Germany people look at me as if they were shouting at me: 'hey there, you stand out; be a grey rose like us, we are all grey roses!' or as if I should not be there, like as if I were a great offense to their eyes. And inside I am like 'no, I am a red rose and if you don't like that, don't look, you don't *have* to look at me!' Of course I don't give them a challenging look; normally I just smile friendly but directly at them.

The thing is that in the society we have created, many people are like this; not just one or two. I am quite sure that each of us displays this kind of behaviour, at some point or other. Yet, if you notice the envy of such a person: be great, be enormous. I am enormous and whatever you (whoever you are) are going to do: it is not going to change me! In other words, stay true to yourself, without becoming hard, nor backing down.

I didn't come to this country to have an argument. I am going to these countries to give something and if you don't want it, just don't take it.

Why can we not just see *I am me*, and that is just fine and *you are you* and that is just fine too!! But where we don't learn to accept ourselves, we will find it very hard to except another and leave this person being a red rose. We will want them to be a grey rose. There are thousands of tiny details like this in our daily life and I really would like to ask you to have a look at yourself and see how wonderful it is to be a friend of a red rose, even if you are a tiny grey little margarita.

And how wonderful it is to be next to a red rose. It is wonderful to be a friend of somebody big. Why not just enjoy it… It's so easy to do. Why do we have to go to the place of envy?

But unfortunately in our daily life we find it very hard to have someone sitting next to us who is being right all the time, while we commence to hear that inner voice whilst starting to scream at us 'you are wrong, you are wrong'.

This is happening *because* finally we are meeting someone who actually is *not* doing wrong. If we want to grow, this can be a wonderful present. Take the red rose how she is, with beautiful petals and without looking only for the thorns.

Now where does the envy come from? From a place of lack, from something inside of ourselves that tells us we are not good enough, and so we don't like others to be good, courageous, different etc. as that to us is a threat, since we don't dare to be different ourselves.

It even has a double effect: first I act this way to make the other person feel a little bad, which comes from a place where I don't feel good with *me*. Doing that in itself makes me feel inferior... that is a double negative effect. (you might have to read these lines a few times to really understand them) By not just feeling this, yet actually putting it out into the world, we create a place that is just not as nice as it was a few moments before.
There are a lot of these 'little' Ego things that we do.

Giving

Let us talk about giving.

There are many ways to give, and often the way people give, comes from habits, customs or patterns.

As everyone celebrates (or does not celebrate but lives) a birthday every year, it is very widespread that, when *your* birthday comes, you invite guests. And it is also very common that these guests bring you presents.
I am sure that all of you are familiar with the fact that you have presents that afterwards get put into a drawer or cabinet as they are not at all what you like. The dead drawer, the dead cabinet; you won't look into this drawer until a next birthday of someone else comes, someone you are not particularly close to, yet where you are invited to and... feel obliged to bring a present. You all know this feeling 'oh God this present, what can I do with it??!!'

When you don't immediately find a person to give it to, one day you find the present in the dead cabinet and you think 'Who gave me this present?!!?' You might think it suits the person of the birthday, but what if it was him/her who had given it to you?

It's better to not even go there!

When you have a present that is not for you, you can say 'it is soo sweet that you thought of me and so sweet that you bring me a present but I must be honest: for me your coming is enough; this present is not for me so I need to give it back to you'.

You might still find this a bit rude or loveless but keep on reading and you will see why that is the better way.

I give you a nice, clear example: A few weeks ago I was in Stuttgart and 2 years previous, the daughter of the lady who hosts me had left 2 orchids with her mother when she emigrated to Israel. Seeing them in her kitchen on the windowsill and another 2 in her living room I asked her: 'Rosemarie, do you like orchids?' Her answer was really interesting as it was a super 'NO, I don't like orchids at all because the main part of the year they are just boring, ugly plants and only a short time they get all the flowers and then they are boring plants again!' Her tone of voice was, while saying this, quite vehement.

She, Rosemarie, is living in a flat on the 7th floor. She can't put the plants in a basement as she would have to water her plants in their boring state as well as in their wonderful state, which would mean she would have to walk all the way up and down all those flights of stairs as she has no lift in her building. Walking the stairs for that is quite an ordeal, especially as she is not a young woman.

Now the hammer comes: A few months later I came again and there were actually more orchids! So I asked: 'Rosemarie, I thought you don't like orchids!!'
'No, I don't but people I invited to my birthday gave me orchids for a present.'
Last time I came she had 7 more orchids so I asked... 'Have you finally started to like orchids?'
'No', she said, 'but my daughter came back from Israel and she brought me more orchids!'
Yet... all the time Rosemarie does not like orchids...
She never said anything about this to anybody except to me because I asked.

What happens when one lives this way: What are you actually doing when you accept these presents?...

You are constantly busy with negative energy.

Most people, of course, are not aware of this, nor of the fear that causes them to be so called polite; out of fear for the, possibly negative, reaction to our: 'I don't like your present'. They are afraid of the possible avalanche that might come back to them. Neither do they realise the energy they are producing, as in Rosemarie's case, that she is generating lots of negative energy during these 2 years. She does not like orchids so every time she looks at them she is looking with dislike; she loathes the people for having given her something she does not like, she loathes her daughter who brings her more, she loathes the fact nobody seems to understand she does not like orchids... In this daily life thing many people are caught up.

Let me give another example: Just think of meeting someone on the street and this somebody gives you a nasty comment and you go home taking it with you... you will still feel uneasy all the way home and even later still. Instead of saying to the person 'This wasn't nice!!' and on you walk; then before you are even home it is gone. When these negative things stay in the system, they start to putrefy, just like food.

When you would be eating solid food all day and then you eat fruit, fruit starts to putrefy. Fruit needs to be digested in the intestine and not at the top of the meal; it will start to ferment and will want to come back up. It will make us feel uncomfortable.

With these kinds of things it works similarly: when we load them on top where it does not feel good, it will start to do its thing=ferment, and come back up and make us feel uncomfortable.

Yet please do not be mistaken: I do not mean to say you should start to spit out at every one who says something to you, you don't like. The idea is to first take it in, whatever it is, feel your emotional response, and without reacting emotionally, say: 'I don't like that'. Because if we just let it sit we are creating karma for both of us and the world is a bit less of a good place. This is not what we want or?... we want this world to become a better place, for all of us!!

If you want to behave in a way that contributes to make this world a better place; not just for you but for everybody; then you are a spiritual person!!

You don't need to be a wandering library of spiritual knowledge.

Karmic Chains

Now if you permit me, I would like to show you a very different situation:

Imagine a family in which grandpa is a lawyer, his son has become a lawyer too, and he has two kids, a son and a daughter. Their father is determined to have them studying law too; he wants them to be lawyers... yet... *she* wants to become a dancer; a professional dancer that is. Her father tells her that that is totally not on! It is wrong to want that! 'What when you are 29? Then you are too old to dance, your career is finished, over!! And then what? Do you think you can find enough students to teach? Will you like teaching? I don't want to hear about it! You are going to university and that is it!!'

The son wants to be a farmer, but his father finds that stupid, a waste of his intellect. So after having listened to dad's sermon to his sister, he desists in protesting and goes to university. He studies 7 years, and during this time his health is slowly deteriorating. Yet... after 7 years he is a lawyer; weak, pale, with ill health, yet he is a lawyer. His former friends don't recognize him when he comes back to the village but his father is proud of him... of the son who gave up his real life for a fake one, the life of his father.

Situations like this one happen many times and they have caused us to have a world with 2 different kinds of people: one kind of which there are many, and one kind of which there are very, very few... I call the first kind the grey roses, and the 2nd kind the red roses.

Now let's go back to the daughter. *She* has not obeyed, she has become a dancer. She is famous for she is a real good dancer; she is happy, radiant, alive.

Her brother is not happy; he has followed the conditioning pattern of society; he has become a grey rose.
She, his sister, has become a red rose; possibly always was...

That does not mean she has it easy. People envy her, some for her talents, some for her use of free will, some for her looks and so on... yet she does not let herself be corrupted by the fear of not being liked by the rest of the world.

Her brother does, he is not happy, his health is poor, he does not live the life he wants to, but... he has no one who envies him, he has a lot of people with who he can frequently meet without major friction yet, to whom he feels he cannot tell what is really his reason for feeling depressed.

Red Roses and Grey Roses

This is the thing you see, that when you decide to be a grey rose, you can move around without any friction, yet you will never really feel alive. When you decide not to live the way you really want to, you will not really live, and therefore you will start to resent the ones that you feel that prevent you from doing this, but also the other ones who *do* live as they wish. You will envy them and not want them to be so 'red', meaning so outstanding, so different from the rest. As you have chosen to be the same as everyone else, you cannot tolerate others to be special. Deep in your heart you want to be special too.

Now do you want to hear a secret?

You *are* special.

Do you want to know why I talk about all this? Deep inside your body and personality is your real self, and you, this real self, is a red rose! Maybe at this moment it is still a bit pink or pinkish. Yet the potential of being a red rose is right there, in every one of you. The real you yearns to come out, to realise itself.

To live up to that full potential you have to start doing what you really want to do, to gather all your courage together and take a step forward saying: 'this is what I want, and I am going for it.' Make what you really want to do, your goal. Nothing else is worth that dedication, not even pleasing your husband, wife or children.

Now of course, being a red rose is quite a challenge. You will be confronted with people who envy you for your radiance, for standing up straight.

Currently this is not bad, it helps you to have a look into what envy is; your own envy too as all people carry envy inside.

Envy is a taboo; one does not talk about envy. People don't mind hearing that they are jealous half as much as that they are envious. They might admit their jealousy but they will find it impossible to admit their envy. Yet, if one can admit this, one is such a different person already, for one can start to look at what envy does, where it comes from, and… if one really has a need to be envious…

Breaking the taboo of talking about it will free you, and that will cause a smile on your face (for which you might be envied again ☺); smiling causes physical and emotional well being.

Chapter 4: Small Things in a Big Context

What is Your Purpose?

Spirituality is not -knowing a lot, spirituality has nothing to do with wanting to become a Zen master; it is not going to a Guru saying 'yes and Amen' or becoming a Guru.

People who understand these facts are the kind of people, the real humanity, who know where they stand in their lives and that they are a big person, an enormously big person.

And if you are not living this bigness, you are not humble: you are arrogant.

Do you know why?

Because you think you know better than the Supreme Energy that has, together with you, created you. I am talking about this energy where you are from, which you are part of. Have you totally forgotten about your service in this world?

We all serve something; we are all good for something in this world. You will have to find out what is yours to do as you are not here just for yourself; that would be very illogical and very selfish. If this Supreme Energy (call it God if you like), and you, would have wanted that, you would have been put on a remote island just by yourself... but that didn't happen. (I don't believe in God as a personification of something so don't think I see him as a big guy with a long beard deciding and punishing... yet God is a good name as we all understand it).

I come back to the word service: you come to this world and you have no idea what you have to learn here and what you need to do here, to give? There is always both and you have decided this 'over there', (where you were before you incarnated) still not incarnated, that you are going to be somebody really big, someone who is going to be outstanding , for example in painting.

And then you come to this world, to this society. You grow up, you have, 'unfortunately', parents, so you become a product of your environment and surroundings and you learn that painting is not good. Painting actually can't help you to earn a living.

Don't paint, become something sensible. Become a teacher or a lawyer if you can or at least start working in a bank.

Many people still think like that, unfortunately.

These people that are only doing a job to earn money, are never really satisfied, which translates into that they never have enough money either, so the person keeps on striving for jobs or improvement within the job to earn more money. The person has started to identify with the values his environment has forced upon him/her. As the real cause for dissatisfaction lies inside, in not doing what he/she really wanted to, there can never be satisfaction however how much money the person earns. Yet when the moment comes that this person gets sacked or goes bankrupt, it might even cause him/her to commit suicide. Money has become his goal in life; the job is the way to achieve this goal…

When you come to this world, you come for something and the idea is to dedicate yourself to that something.

I came to help humanity, which means you. If I don't dedicate myself to it I will be feeling lousy. As soon as I don't like it any more I will do something else. I love to sit here and write this book for you. I love to stand in front of people and talk to them. From the moment I don't feel like it any more, I will do this work in a different way so that the passion in my work remains. I have the feeling that, if I would want to do something else, I could make my living with that too, if it would be something I would do with the same passion. I am sure, that, when we are doing the right thing, all that we need comes to us. This is not just a feeling. This is my experience in my whole life.

It doesn't matter what you do and it really doesn't matter whether you make money with it. Every day we make the mistaken thinking that we need money to live. What do we need to live?! A roof over our head, clothes to wear, food to eat, fire to warm us, people around us to love and who love us, and peace.

That's what we need.

If you have all that and you can really do what you like, work which makes you happy, why would you want to get more?!

But what the society tells you is that you need more money. So when someone else has more money, he puts it in a bank and he has interest. Isn't this ridiculous?! To me it is absolutely insane. What reason would there be to create more money without doing anything for it? If *I* get it, who must be deprived of it?

So we can't live properly, because we created a system that doesn't work.

By the way, in a society where everybody has the possibility to live a life with dignity, all layers of society have a better life, even the very rich. There will be less criminality as there is no need for it... All will have enough to eat, enough for clothes, for a roof over their heads, a hearth for warmth and an occupation they feel good about.

Being Spiritual

We are the system and we forget that. The system is not something over there, the system is you and me and all the rest of us. Yet we created a system in which we learn to lie, to not tell truth. We created a system in which it is almost impossible to survive if we don't cheat the laws we have created...

If we change in our daily life tiny little bits, for example taking on this attitude of actually saying what we mean to say and meaning what we say, we are already changing this.

Because then we are not something different from what we are saying, because if we can't speak clearly and truthfully, it would be better not to speak as we are not *being* what we really *are*.

Check when you speak: When I am going on the right path and being frank, why am I saying this to this person?!

If I say something to you I must mean it, yet I must not say something with an underlying message that I am not actually saying in words; here I should say what I mean or shut up altogether.

If I say only half of the message it means that I want to manipulate. Don't manipulate, say: 'I don't like what you are doing' or whatever the underlying message was you did not say. There is nothing wrong in disliking something.

Just let us decide to not dislike the other person straight away, to be more neutral. That doesn't mean that you like everybody. You don't have to like everybody… or would you like to like everybody?

That doesn't mean your disliking is not good. That is the difference in this story.

There is not a good or bad in liking or disliking.
If somebody is the person who is disliked it doesn't say anything about this person at all. Is just says something about yourself, about you, the person who dislikes.
The chemistry between you and that person doesn't work for you.

That is what it says. It says nothing more. If I said 'I don't like what you are doing now' it doesn't say anything about you. It says something about me! It says that *I* don't like something, not that what you are doing is wrong.

This happens with many things, with these things that we are living every day. We are just projecting, projecting and projecting sooo often.

And we forget that actually all that we are saying and all we are thinking usually says something about *us*. So you might stop talking.

That's very often case, that it would be better to just shut up. Step back and really, when you want to say something to somebody, try to feel what you are going to say, feel inside how the look on your face looks with it.

What is happening inside you… If I feel anger in myself I must not have a smile on my face because the person in front of me is going to be very confused and this person can close up like a wall because what is coming is not real.

Then I must try to explain that I feel anger, yet have no anger emotion in my words. It is a matter of trusting that person to be evolved enough to be able to receive my real self, and not just a projection of the product my personality has become over the years. This way we will create trust between us. We must relearn to trust, that we are not the only ones that understand...

We can start to describe being spiritual as doing what we really feel passionate about and living correctly at the same time, as that is being of service.

Energy is Only One!

We talked about judging someone.

Yet there is something even more important. This judging we talked about is still in direct contact right?

But if I am actually talking badly *about* somebody it goes to the person I am talking about, through underground pathways, even if the person is not present.

This happens because we are *not different* energies. I am not one energy, you are not another energy... we all are just Energy. Just in the way it manifests, there is a difference. But we are all the same Energy, which means if *I* do something it vibrates; it even reaches the person although I don't want it to.

Let me give you a parallel example: our body is an organism, having many different parts, each with their duty and purpose.

So is humanity; a large organism with each person as a part of this whole, having a different duty and purpose.

Now we might find that someone can be disposed off because she/he is a murderer. Let me compare this person with the last part of my little finger. If I would cut this part by accident, most people would say: 'no problem, it is not such an important part, you can do without that'.
Yet what they don't know is that the beginning point of the meridian of the small intestine has its place here, as well as the end point of the heart meridian. The cutting might cause disturbances in these 2 organs because of that. As each meridian is also connected with psychological processes, we will become off balance within as well.

We don't know what the use of the murderer is, yet the act of murdering might open many people's eyes to how necessary it is to change things in our system; to not judge other people; to not hurt, manipulate or exploit someone, even with words, as it can cause devastating effects.

When I cut off the top of my little finger, I will never be the same again. When we dispose of the murderer, humanity will never be the same again... (don't try to be funny by saying that that is exactly what you hope ☺.)

This means that we must not talk about someone negatively when they are not there, as this energy goes to them. Nor must we say things about someone that we don't want the person to know, as it will have its negative effect anyhow. So if we will be doing these simple things that we encounter every day correctly, we are living in a spiritual way.

Now we are back to where we started: living our daily life and making it into a spiritual way of living, is something we do with very simple things.

If you want to be a spiritual person you have to realise that *it*, life, is not about you, this means you will do whatever you can for you first, so you can 'drag' everyone else with you towards happiness, heaven, nirvana, or whatever you would like to call it.

If you are seeking any of this just for yourself, you happen to have fallen into one of the biggest spiritual traps that exist. If you seek enlightenment for yourself for the sake of being enlightened, you will actually never reach it.

Yet, if you try to be the best person possible, aspiring correct living just because you want to live a life of love, you will become happy; yet only if you do it for the sake of being good, without trying to resolve your Karma or trying to get anything else out of it. And in this unselfish way, feeling well, being happy, you will make everybody else feel well also.

The unfortunate truth is, that more than half of the so-called spiritual people think that they are more important than the ones who are (from their viewpoint) not spiritual. They have an enormous Ego. On top of it they need to feed this ego all the time, by giving it lots of attention. If you really want to do something for this world, make sure it is not about you; do it selflessly, so you don't fall into this ugly spiritual trap. Only that way will it work, whatever it is you do. Now do realise: it is important to look at yourself, to check upon yourself, yet even if you do things wrong, make mistakes, do not whip yourself for it!! That is not being spiritual either. Give yourself a chance; be tolerant with yourself without being too permissive.

Of course the same counts for how we behave towards other people. Let people be free to choose their way and give them a chance. Let us try to work together for a better world, not against each other. Try to see the best in people and work with that, and put the differences to the background. Finding common ground, will make the change manifest.

It is all about doing things we like; it is about being a red rose. If we are doing things from the place of love, we can't go wrong.

Don't even take ego as something bad or wrong: ego is the instrument we have been given, to find out where we must change. It is wonderful to see that by feeling our ego, we realise where we are in fact acting from patterns of the past, instead of living in the now; while living in the now we don't need any of these 'games'. If we use our ego to see what it is we have done wrong, we can change it and won't have to prolong this pattern.

Realise that negative thoughts are very human.
Yet, it is also human to do things using our willpower. So we can use our willpower for positive thinking and this is not only a chance; it is important. It changes the life of the person practising this as well as the lives of those around this person.

If you allow me to give you advice how to set to work with this I can give you the following ideas: start with those qualities in your character where you are quite good and try to do that even better. You will see your progress and feel more like taking on the next task.

I would not recommend starting with the things that you experience as bad and difficult, as you might lose faith in your own capacity and fall back into the old patterns; it could even cause you to fall into depression.

So if you are a fairly honest person but not always, you can start to improve there, becoming honest in each and every aspect of your life, and automatically you will become more aware that you might be manipulating, envious, doing things without love and so these things improve because of you trying to be more honest.

One day you feel you really are as honest as you can get and you might try the envy: you will try to become more generous so there is no room for envy etc.

We are going to change the world bit by bit. Whatever you are doing will have a snowball effect. We will not start building the house at the roof. So be patient with yourself.

You want an example of someone being a living example?

Since several years we are going to a particular garage when we have problems with our car. The owner is a very nice man called Paco. My daughter had some problems with her car and needed to go to this garage. She has quite a temperament. When she is happy, she shows lots of burbling happiness; when she is not she exudes thunder clouds.

When she came back home after leaving her car at the garage she exclaimed: 'Oohh Mum, this man is unbelievable because when I arrived at the garage I was really angry and only after a few minutes talking to him (Paco) I came back like if I had had a therapy session.'

Paco is a person who never raises his voice, never has anger or contempt in his voice, yet always, very calmly and quietly, conveys what he has to say. Through his way of calm he always gives one the feeling of wanting to agree, or at least come to a place where both agree. The example of his *being* is having such an impact and positive effect; not just on my daughter.

That *is* what happens.

We often forget that we really need other people. If we were isolated, totally alone, we would be very unhappy.

As soon as we are capable of admitting this fact we can start to improve our ways with others. We realise that we make mistakes and can ask forgiveness, apologize. If I really want to change the world I have to start within, with myself. Michael Jackson sings it so beautiful in 'the man in the mirror': *I am talking to the man in the mirror; I'm asking him to change his ways; I'm talking to the man in the mirror: if you wanna change the world into a better place, take a look at yourself and change your ways.*

We can start the day saying into the mirror: 'hi there' (put your name here for example: John) 'hi there John, I am great!' and smile to myself. This makes me feel good, so when I go downstairs to meet with my family with this feeling that life is good and here and now is where I am, my situation is good.

What does this do to my family? This causes my family to feel that they have the freedom to feel great and do what they like and be happy. Then the chain or snowball effect starts as they go into the world and pass these feelings on too.

If you are a person who doesn't need people at all you are either perfectly in unison with All that Is, or you don't have a clue, which means you are closed inside your prison made of patterns and circumstances.

Now let me come back to the red and grey roses: as soon as I take this stand of being the red rose that I am, not making myself a pink rose, but really strongly be the red rose, in that moment I am giving everybody else the freedom to be a big red rose.

I am radiating to you: 'you see me being colourful, be colourful too!! Communicate in your colourfulness'. I am actually telling you: 'hey, come on, be who you are!! Show me what there is in you, there is such a humongous potential.'

Yet when I am sitting in the corner pretending to be a little grey mouse I convey to you that you have to be a little grey mouse to, or at least behave like one.

When we live *this* way, the grey mouse way, become part of some educational system, family, school or any spiritual group we are not going to make any difference; it will be as if we were not there. And that is the best of all possibilities, because actively it can happen that this example is followed up by others and we corrupt our society by being 'grey'.

Being an Example

Yet if I am being a red rose I am showing that that is what you can be too, what my students can be too!! And let me tell you: the red roses don't envy another red rose for being one; they admire them! There is plenty of room in this big garden for all the red roses to be…

Yet, if the way I live is showing inhibition, dishonesty; if I manipulate and smile at you with a false smile, I am showing you that that is the way I find that one must live, this is what I will be radiating and people will, unfortunately, follow this behaviour as an example. If we have a good example in front of us, we will follow this and behave alike. Yet if we have a bad example near, we follow this bad example and accordingly become less nice people.

When I give conferences, I often, without telling this, start to rest my head on my hand; within 10 min. most people are sitting like this. Then I will start to scratch my head and also within 10 min. most people have scratched their heads.

This has been even so intriguing that there are behavioural studies about these things; they test people with very simple things, and the grey roses follow, the red roses don't. They just observe and stay themselves. It truly is imperative to be a good example, and when you are reading this book you have already come out of being a grey rose; you are already being an example somewhere. Yet realise, you ALWAYS are an example, so now deliberately be a good exemplar.

That is how life works: people learn by copying. This happens mostly unconsciously, yet it happens; people follow models. The more good examples there are to follow, the quicker we will change the world. Here you can see how important you are. Even just you changing will transform the people around you. YOU MATTER!

What is success? Success for me is if at least one of you reading this book will put it into practice, putting this out there where people will take the example and follow it. And super success is when all of you would be doing this. And already several people have given me feedback of how much this book guides and helps them every day.

So stay there being a red rose, even if you get no positive feedback. As a red rose you must try to stay in your centre, don't change your level of vibrations, don't lower it as that will only attract disaster. If you get tested for example by being treated disrespectful, take your time, maybe you find out you don't even need to react. Don't put yourself under the pressure of having to react instantly.

Realise that if *you* are being the red rose you are, you're showing all the others that *they* have the allowance, the freedom, to also be the red rose *they* are and you inspire them to be courageous.

Every little step you take and succeed in taking, matters! Your path is your goal, your destination is not your goal as your destination is not fixed; it does not exist. The only destination we have is change.

Everything is now, so now matters. Look into this world with interest, show you don't know everything, because you *don't* know everything and there is nothing wrong with that!

Be aware that sometimes what you like, might not be what you really need. Life is what it is and we have to work with exactly that: with what is. We are not perfect, yet what we have we have, what we are we are right now.
Yet everything can change.
Don't try to work with the idea of what something can be, as you don't know if it is going to be like that. You would be working for nothing as what you want it to be is an illusion.
Work with now, with what you have and live now. 'If you can't be with the one you love, love the one you're with'.

I can hear you asking: 'but don't we need to make plans?' Of course we need to make plans, because a plan gives us the idea where to start. But a plan should only be a guideline, and we must learn to be flexible and maybe discard the whole plan after only one step...

Chapter 5: Rivalry

This is a really interesting topic, because rivalry has been such a wonderful theme for films and books. We see two comical figures as rivals, or the detective and his assistant, the knights fighting for their honour or for a lady, two women wanting the same man etc. Yet that is not really the way I am going to talk about it.

I would like to start with asking you several questions, and please, when you read them, answer them honestly. It will help you to get to places in yourself that you can clean and clear and afterwards, not only feel happier, but also a lot more relaxed…

Are you a person that finds it important to be strong?

Is it important to you to give the impression of a strong being?

Are you often tired, yet feel that you must work more?

Do you like to mention the fact that you work really hard?

Have you boasted about how much you can drink?

Now if you have answered any of these questions positively this is going to be very interesting for you. Yet, even if you haven't, you will get a lot out of it, either for yourself, or because in this you will recognize people who you know.

I know this lovely woman who works with horses. Let's call her Sue. Working with horses can be really hard work. And her work is hard: she cleans the stables early in the morning: her day starts at 7.00 am, and then she feeds all horses, normally around 14 of them.
Recently she has started to sit down after this first chore to have breakfast, which is after 2 hours work.
She then goes back to the horses and starts to either ride them, walk with them or do groundwork with them. These are raw horses; at least they were when they came to her, so there will be horses at all stages of first human contact and training.
If you know anything at all about horses: they are very strong and can walk/run fast, so the work with them is physically hard and a lot of it is running… in this first training phase.

Back to Sue: in between the training of horses she will again and again clean out a stable or a paddock, as horses happen to drop excrements during the day. By the time she has ridden 7 to 8 horses, it is 8.00 pm. She might have had a short break somewhere in the afternoon to have a banana or even a coffee with a visitor, yet she normally works more or less non-stop. At 8.00 pm it is feeding time again (of course at 2.00 pm it was feeding time too) and she checks if the ones that have no automatic drinking-device have enough water. Normally they don't which means she fills up 20 litre cans and carts 4 of these in a wheelbarrow to the different horses.

Sue is 43 years old and has done this for almost 20 years! She does not call this her work though. Her real work (that is how she phrases it) is service... this means she works for a catering service where she prepares the place (setting tables, decorating etc.) and serves food and drinks, sometimes even 24 hours nonstop, and afterwards still has to do the cleaning up and gathering the rests... this includes carrying heavy trays and boxes as well as crates with bottles. You can see that it would not be bad if she had a male assistant with a few healthy, strong muscles! But she does this all alone. After having had a weekend like this she does not rest for a day, she goes to the horses immediately.

This is not the point though. The point comes now: whenever there is a party where men are present, she will drink them all under the table (or at least almost all) and is proud of it. She is also proud that she can drink lots of wine and then harder stuff like gin or liquor and has no problem with hangovers the next day. She drives her car like a formula 1 racer and is proud of this too. She won't easily ask for help while having to carry heavy stuff as she can do it as well as any man. When she does her exercise in the morning, she does it heavily with lots of strain and pressure... when going out with others on a bicycle she will be the one going fastest and on top of it going over fields where the surface is difficult, showing she is strong. She has to give her opinion about almost everything, and becomes quite stern in a patronizing way when she wants to show someone she is in competition with, that she is right. Now this gives the clue: she is in competition....

When we look where she was born and what happened, things become clearer: she was born as a girl and her father made it very clear that she should have been a boy. She has been trying to be a boy all her life. It has caused her to be in competition with both the female and the male gender, to the point that she, with all her womanliness (I purposely don't say femininity as there she is progressing only slowly) will have in her handbag many things that men like. She is super proud of the fact that, in a circle of men all interested in the technical side of cars, one will ask for a folding-rule, and she is the one who has that in her handbag… She needs to be accepted by men as a man so she is in rivalry with them, and that has become a way of life. She is in rivalry, not just with men but with most people she considers strong and… most of all… with herself! She is constantly proving she can be better, harder, tougher…

All people around her know she can do sooooooo much and no one finds it necessary that she should prove it. Yet she does. This goes on for many, many years. She is not getting younger, so it is taking its toll. She is getting more tired. This is something she would never admit, yet after having worked together with me now for about 6 years, she sees that being tired and being strong don't necessarily contradict each other.
She has now come to a point, where she is actually hurting herself. Not on purpose; her body is simply rebelling against overload, so she will hurt a vertebra, or her shoulder will not want to go down after heavy lifting; a nerve gets stuck somewhere etc.

Why do I tell you this story?

Well, if a person enters in such a dynamic, she/he will start to become pretty hard, even harsh, in the way she/he speaks. This will cause discontent in others, also in people they find important.
They find it hard to loose, so they will not easily admit and become very stubborn. This goes so far that when someone else has done something wrong to them, for example caused them some sort of pain or grief, they might break the contact with that person for ever: the door gets thoroughly closed.

The real reason for closing this door is that, if they would allow that door to stay open, they would have to look inside of themselves. There they will find what they will call weakness, and weakness they cannot allow (yet). They will see that they are vulnerable, that they hurt, and that is not something they can effort (they think). They do not allow themselves to be mirrored, as they have to be flawless, free of failure. You can imagine what a strain that puts on them as no person can ever become that perfect. So yes, they are perfectionists, but don't be mistaken: not all perfectionists are rivals. A sheep is an animal, but an animal is not necessarily a sheep.

When they start to learn that making a mistake is human, that saying sorry afterwards shows enormous strength, especially strength of character, they have a chance to come down from this high and quite lonely mountain they placed themselves on. They will start to relax and get feedback in a gentle, soft way. The patting on the shoulder of comrades will change into a warm and loving hug... but 'hey, slowly!!' they cannot hug easily, don't press too much, give them space to see that human closeness is not something to be afraid of.

And here we see another important part of this: they find it very hard to let human closeness into their lives. Often they are people who love dogs or horses, particularly these two kinds of animals, as they fear the independence of beings... dogs and horses when not in the wild need human company and care.

They are usually very good in helping others. They love to give others what they won't give themselves, as that is showing their strength. Yet accepting help for them is very difficult. They often even react as if the person who offered help were treating them in a patronizing way, as that is how they perceive it.

If we would all realize that we have been born as a human being, amongst other human beings, and see that we are all interdependent, we will not have this dread of closeness, asking for help, not being perfect and similar situations any more.

When we have grown up in circumstances in which we were feeling that the way we are was never as good as the way others are, we will develop this strange kind of competition with the world. We will do everything we can do be better and in the end we go so far, that we want to become better than ourselves. You can laugh about this, yet this unfortunately is true far too often!!

The person who behaves this way will look like someone very adequate, as she/he has trained her/himself to be as good as is possible. They are normally really good in their profession.
In brief contacts they are successful, people will really like them, yet in long standing relationships they have problems as only those, who see through their attitude, can appreciate the heart there is within. Because the challenge of being with them in the hope they open up, is almost constant. One can trust them, yet they find it very difficult to open up, even to those they trust. Often they think, that their problems are too much for others to deal with, as they have it difficult enough already.
It is not easy to get very close to them; they need time and space. One must not overwhelm them, but give them little 'homeopathic' doses of love, care, information and opinion or observations about themselves. They will think about everything if one does this carefully, as they want to be perfect (don't forget that!) and will slowly see that their lives can improve if they let some of the imperfection in…

People, who don't treat these auto-competitors like this, will not like these competitive people as they will take it personally and don't see that this is a pattern in the other, not having anything to do with *them*. Often they find them too harsh, to judgmental, and at the same time they get envious of their (superficial) popularity and strength.

Do you recognize some of this? Do you know people like this?

If you do, even if the yes-answer is to the 2nd question, then have a look into yourself how you could get out of this rivalry with yourself and the world. It will make you more relaxed, more loving towards yourself, softer towards others and as a consequence of this, others will be softer and more loving to you; you will not even feel compelled any more to compete or be loud. Of course, this will take a while, yet you will see that you will be making progress gradually and that you will be enjoying that progress all the way!!

Chapter 6: The Dark Forces

I have been talking to you about what we do in our daily life. It has led me to talk about the negative qualities that we have in our characters. Not everything that we do wrong is just our own doing. We are influenced by many forces, yet we can get stronger so that we will not be influenced by these forces. I will talk about this in this chapter and in the next.

In creation a lot has happened that we, as 'simple' human beings, would not have sought to create, had we had the choice. Some of these 'things' undoubtedly are caused by the dark forces. Most people think we would be better off without them. Yet, how can we really evolve if there are no challenges… That does not mean that you start to like or have to start liking the dark forces. Yet try to see them as a way of making your inner growth more thorough, more durable, definite and stable. So when we talk about growing on our spiritual path, we must also talk about the dark forces.

Ahriman and Lucifer

About a hundred years ago, Rudolf Steiner warned us against the dark forces and depicted them as Ahriman and Lucifer. I will not enter into the specifics, but he notified the world that Ahriman would incarnate into our world to make life hard and difficult, and this would most likely happen around 1984, but it did not and Ahriman is still preparing. Lucifer is a fallen archangel, so incarnation will never happen. Yet Ahriman can and, according to Steiner, will start to bond with Lucifer to make his work more powerful and prepare his incarnation thoroughly. We were lucky that he did not incarnate, at least not yet! Yet Rudolf Steiner did not just tell us a fairytale; Ahriman and Lucifer do exist and their forces are very present. Yet what I am going to say is not to scare the life out of you, it is there to help you to become aware of where you can improve your life.

Now as *I* am writing this book, I am not going to repeat something someone else has said. I will speak from my own vision and experience and will explain the work of these dark forces the way I see, live and have lived them.

Let me first tell you how these two, Ahriman and Lucifer, manifest the way we can experience them; what their goal is.

Ahriman wants us to become will-less, robots. He likes to make us into slaves and work for him without using our individual will. For that he created things in matter, as the material world makes us greedy, so we want more of it. He created gameboys, computers, mobile phones, televisions, the whole internet, as to have us spending lots of time sitting behind machines and struggling with them, in the wrong conviction that we cannot do without them anymore, and many people cannot let go of using these things and ignore the effects they have on them. Before being enslaved they would do things like spending time in nature, spending time with friends and families, shop in the local shops instead of using internet shopping possibilities which in the end are not good for our environment as they use lots of material to wrap up their products and the transport from the most remote places for such a product to the home of the buyer is also irresponsible thinking about the environment.

But, what I think you all know most: the time you are wasting when at your computer because it never goes as smooth as you thought; it never is: just 5 minutes to check my emails (and before you know an hour has passed). Just think of all the links you have to follow etc.

Anything that makes you focus on just something that takes your attention away from what you really want to do in your life, is the work of Ahriman. It also shows in those sensations where we feel inadequate, where we get the idea that we are far too small to make a difference, and forget to listen to our inner self <or some people might call it to our soul> and forget to see that we are on the right track.

Ahriman tries to eliminate anything that has to do with spiritual science and research, and endeavours to eliminate our willpower and the use of our free will, which we can see in the way how all of a sudden many companies (mal)treat their employees and how the states take decisions in spite of their citizens not wanting them at all, and how few of these citizens then really protest openly and try to create change.

He wants to make us dependent, so he inspired the inventions depending on oil so we need to find new sources of energy as oil is finite, and all other forms of strong, easy accessible ways of generating electricity are also, even possibly dangerous (nuclear power). This makes us worry which is, together with making us dependent, the ideal combination of keeping us far away from doing the real important things.

We have become slaves of our material needs. Of course in all this it is important that there is control over us while this is in process, so there he inspired the creation of what we have now as the so-called social networks: facebook, twitter, linkedin etc.

These are also under the influence of Lucifer who tries to show us how handy and important it is to be part of these networks… I will talk about Lucifer in a little while.

Ahriman also instigates the way of thinking in frames. So many people will only be able to accept something when it is presented in a logical framework. Yet the framework is fixed, the reality of *growth* within evolution is not taken into account and therefore not lived and nurtured. Can you see how dangerous this is? When later you read the part about the influence of Lucifer and I now tell you what their working together causes, you will understand this even more. Because under the influence of the two there are many people who stop developing their own talents and learn constantly more new techniques. Yet each technique is limited, as it is developed by one or a few people.
That is why my therapists are all different, as their training is not a method, but a non-methodical method, designed to develop each person's qualities.

Lucifer is very different, Lucifer makes the wrong things look nice, so we don't see they are not the right ones. For example, he helped create all those gadgets like beautiful candles, shawls, oil lamps, meditation cushions, mantra pictures, spirals and machines in colours and shiny materials as to have you believe you need these to become enlightened. And if you have enough of this stuff, your ego will grow and you will believe you are actually more spiritually evolved than all those who don't have these things.
He will make you go to seminars where a lot of kissing and hugging and emotional expression is going on, as that looks and feels good, yet has no future effect and will not help strengthening the real force in you, only your ego. He will show you that a seminar where you can dance, shout and let out all your emotions in a very emotional way is the thing to go for.

Yet these seminars don't bring you any further as the emotion is not even conscious and the root of what you can discover about yourself has not been revealed, but you have been made dependent on these seminars as, like any drug, for a short while they make you feel good. Lucifer will make things look nice, also in the behaviour of people, who, without anyone noticing, can manipulate while under Lucifer's influence.

Lucifer feels like a snake under the grass, yet most people have no active experience with this phenomenon; therefore Lucifer is very successful as most people are totally oblivious to his workings.

At the moment we are living in a time in which these two forces have bonded. This is only since around the year 2000 when this started to develop. That makes it so much more difficult for us humans to deal with, as all the slave making is wrapped up in beautiful looking possessions, so we fall for it and become will-less. We will want a more beautiful couch or a new pair of shoes of the latest fashion, in spite of having a comfortable couch in the house and plenty of shoes in the cupboard. We will spend lots of energy in achieving, wanting, looking good, having the right furniture and car etc. instead of growing in human relationships.

What we can do against all this is to reinforce our openness to inspiration and intuition. That is why I created **The Course**, so that people learn a new, independent way of judging and think for themselves, despite others trying to influence them; that they learn to see where someone, or they themselves, might be manipulating to put a stop to this immediately. To learn to become disciplined so the willpower gets reinforced and to become clairvoyant, developing the intuition, are important objectives of **The Course**. From here we will be able to find our willpower again and actually *do* good.

When one is very consciously working with and for the Light Force, it does not mean one is not vulnerable to these dark forces. On the contrary: we who are dedicated to the Light will get attacked by them more than those who are not as these forces want to lame us, eliminate our effect and influence on the world; make it impossible for us to help others to shine their own Light.

There are many ways they do this, I will tell you about my personal experience.

Elementary Beings

Ever since I was born into this world I have had a very good connection with the elementary beings. I see the little helpers all of us have around us, the little ones that look after trees, flowers and plants, the beings in the wind and the water etc. I see them and connect with them and always loved the communication with them. Some of them have shapes like human forms, often far smaller, others look very different and not like a body anywhere near human shape. They are all light-beings, some denser, some lighter, some long and thin, some short and stubby, some so transparent it is hard to see them, some clearly colourful.

Physical Discomfort

During many years I have given conferences, seminars and have also given private treatments to help people to live their lives to their full potential and to recuperate their health. At some point my own health started to become a problem. Every time this happened I worked with it and it improved. Then it would become worse again. I went inside myself, looked at the cause and again could improve it.
I had a severe intoxication in which my kidneys totally stopped working, yet I achieved the impossible and revived from death. It took me a long time to become strong again, physically, yet I recovered in a way that no one ever dreamed of.

Eight years later I broke my neck (1st, 4th, 5th and 6th vertebra; the last three had fused together like a telescope and the 1st one had a piece broken off that had moved to the right) and the accident also tore and moved my two right cranial plates (because the blow had shifted my jaw). Before going to hospital I set this in order as I was afraid they would open my skull and would cause haemorrhages and I would bleed to death.

The hospital almost let me die, not noticing anything broken, nor the seriousness of the situation so they sent me home again; the specialist who saw me after 10 days without getting the right treatment was aghast when he saw how I had (not) been treated and understood I was not going to stay in hospital. I set my own vertebras as well as the cranial plates... Also this time this was considered to be a miracle and the normal medical world was stunned.

In this one accident there were around 12 medical reasons to be dead, yet I wasn't. I did not even stop working. The accident happened the 15th of March and on the 3rd of April I was on a plane on my way to Germany, where I had a seminar scheduled. I could not walk, so I was in a wheel chair as I was partially lame, yet I gave my seminar in a comfortable deckchair and no one had the feeling that I was less present than at other times.

In the mean time I realized how thoroughly tested I was, yet... I went on. However then my health started to deteriorate in a different way; I was getting very tired; there were many things I could not eat as I would have a bodily reaction of pain to it; my ears started to get strange, my eyesight also, not less hearing and sight, but things like pain, stings, defocusing, high pitched sounds or ringing sounds and so on. It became more and more distressing, especially because I was doing the right diet for liver, kidneys, pancreas, stomach etc.; lots of good and soft exercise, enough sleeping, eliminating stress factors from my life, and yet... lots of my attention had to go into this feeling of physical discomfort. I looked and looked and looked inside of myself and could not find *a* or *the* cause. It came to a point where I felt I was stuck.

Desperation

This is something unusual for me, as only once in my life I have had the feeling of being stuck and that was a long time ago, I was 23.

This was really tough. I felt quite desperate and told my husband, who, although wonderful as he is, could of course not help me.

I thought of calling Sharon, my dearest sister-friend, in Canada. Yet I resisted, having the feeling I should be able to do this alone. Bingo, here I got it: for me asking for help is far more difficult than doing it alone... And I saw the trap, or part of the trap, I was in: this is not right: I don't have to do it alone, I can ask for help!!!

The first hurdle was taken. I went to the back terrace of our house with my cell phone and called her. I told Sharon my whole experience; especially that I am getting so sick of having to put almost all of my attention on my physical body, leaving little time and strength for that part of my life I find important or even just nice. She listened and then just said: 'my feeling is that what you told us years ago about Ahriman is happening; he is trying to get you out of his way so your light does not shine on us and the world! Try to listen to those little helpers you told me about, as they will tell you what you can do.'

At that moment I left the terrace and stepped into the garden and connected with the elementary beings, tears running down my cheeks. I cried a lot, yet all the time was in unison with the elementary beings; the ones that help me personally as well as the ones in the flora of our land; the ones beyond that, the wind beings and all that I could connect with, while still talking with Sharon on the phone.

It was enormous what happened: they really rejoiced!! It was so strong what happened, as for me the contact with them has always been so natural, yet I had never felt that they needed me as much as I need them. That without me being a mediator between both worlds they cannot do their work...

Isolation

The fact I had been cut off from them was because Ahriman was trying to isolate me, and he had almost succeeded. For a year I had not actively been in communication with them. What happened then you might not believe, yet it did: the plants and trees around me attained more colour; the grey veil that I had seen over them for the last few weeks slowly started to lift and I could actually see how it lifted like a wave from the whole valley and mountains around me. Not fast, yet very steadily. Bit by bit the green became greener, the red redder etc. I was very moved and grateful, to them, to Sharon, and to the fact that I still always am capable to find the place within me, where I just am and know the next step to take. Yet this is not the whole story.

A couple of weeks after that we went to Germany where, during the time I am not on tour for my work, we live at a friend's house. When we arrived, stopping in the middle of the road and were surrounded by friends who welcomed us, a woman I have treated in the past, made a point of inviting these people, these friends, to a horse riding event she organized and when I said: 'Oh how lovely, can we join?' she very bitterly said no, they wanted to keep it close company. Yet the 'they' was just her as all the others who were invited were our friends.

She made a big point of wanting to isolate me from this group of friends. She knew that all these friends would go and no one even spoke up as to plead for us to be able to join, as she was the organiser. At first it was painful and I was upset, until suddenly I realized how similar this was to my previous experience: Ahriman wants to isolate me, so I cannot do the work.

A week later another situation: I promised to make photos of the people doing a horse riding seminar and, as a friend who was staying with us, would also go to the event with his car, he would take me back and forth, so as not to use two cars and have two cars polluting the atmosphere.

So of course the deal was that I would return home with him as well. When after the event he left to drive home, he never told me: he just let me stand there and I had to walk.

As I was exhausted from the whole day on my feet taking photos, I was exasperated as it was quite a long distance. On top of it I was wearing shoes not suitable for long hikes. I cried of tiredness and desperation and prayed to 'God' to help me to deal with this guy the right way; not from my emotional discomfort.

When I asked this man why he had not taken me back he lied, and this was obvious to everyone. I told him this, without emotion. The other people in the house, who observed this, never said a word; I was left alone in this. Again I realized: the same tactics; Ahriman wants to isolate me and make me feel hurt so I have no strength for what I really want to do. Yet he could not get me as I was not resentful: I said what needed to be said and left it.

Now the most important disclosure was that I understood about how he works: *he wanted me to fight him.*

I did not. I just turned my back on him on all three occasions; then he has *no* opponent. As an opponent I would recognize him as a disturbance and give him power. I ignore him, except for informing others about how he works and what they can do.

Yet right after the situation in which I called Sharon, I could not talk about it, as I felt I was not strong enough *not* to give him power while talking about him, as I was not sure whether I could be emotionally unattached, and so I asked Sharon if she could tell my daughter Erinda about all this.

Erinda is a very aware young woman, and if she knows about this situation, she will be prepared to identify where he is at work in her world. I want to share the email Sharon sent her with you:

Dearest Erinda

Your mum has asked me to describe the process she has just recently been through, as I was with her in it, and am more ready to put it in words than she is. I feel it's important for two reasons.... one so that you know where she is at and how she is feeling, and two because this is something that concerns all of us humans.

She has been struggling so long with her physical symptoms, and with worries that get her down. Worries about her health, about being able to make the difference she really wants to make in the world, about doing it right, about being able to discern whether she is doing it right or not...... and so much more. This struggle and worry had exhausted and weakened her, using up most of her daily portion of energy and leaving little energy left over for anything else. When she phoned me on Sunday she had reached her limit, and was crying so much.

As she spoke, and cried, it was like she was crying not only from the pain and frustration, but also from loss. And it was clear then that what she was missing was her support, her family, of higher beings. Here on earth she has us, who love and support her in all ways we can. But she lives on many levels, and on the higher level I could see that the support system was there, but she wasn't counting on it and interacting with it enough to have the strength and guidance to do all she has come to do.

And we realized that Ahriman has been busy there. He is subtle... as your mum quoting Steiner said: "he gets under the skin" He is the ultimate abuser.... abusers isolate their victims, and lead them to believe that it is all their fault. Through emphasizing the worries and the pains, Ahriman engaged your mum so much that she paid more attention to them than to her higher companions.

This way she lost her sense of peace, and he could then create more of a rift between her and her higher support system. And in feeling alone, she could blame herself for so much. And this weakens even more. The higher a person is in the hierarchy, the more important they are to the larger plan. Because your mum is so important, Ahriman has been hard at work to keep her from her destiny.

But we won't let that happen. She realized that her spirit companions are always there, and have been waiting for her to turn to them and interact consciously and continuously with them. To do this most effectively, she had to turn her back on Ahriman. Literally turn around. Put all her attention on the higher influences. She herself said this will take much attention and focus, as Ahriman is a strong and sneaky opponent, and he is making a huge effort to keep her away. To not look at him, to deny him acknowledgement, takes away his power.
This is not the same as to be in denial. We must be very clear of the danger he really is, at the same time as being clear that he is truly a waste of time and energy. There are better things to do. So now her energies and efforts are turned to hearing and speaking with those that are working with the light in the higher realms, instead of looking at Ahriman. Speaking with them is coming easier than hearing them, so she is taking time to be alone and allow that to happen.

Her grief and sense of loss is much less now that she feels embraced again by those that truly support her. In fact we both noticed how everything changed when she moved her focus to these beings. They opened their arms to her, they were relieved to have her back, they rejoiced.

Even the colours of the trees and light that your mum was looking at changed, became more intense. I felt that the universe sighed with relief and anticipation; because now they can get on with what they really want to be doing... Ria is with them.

It works both ways; we need the support and guidance of these higher beings, and they need our intent and interaction. It only works when we are together.

I love Ria so much. I love you so much. My dearest, dearest family.

Big hug, Sharon

Sharon Loerzer, Sa'Sen Yin Therapist

After this Erinda realised, that with her Ahriman used her fear of not having enough money when she starts to study again as well as her wanting to be a really good friend and feeling inadequate there sometimes. He had her there, yet not any more. She sees she can trust the universe that she gets what she needs, and being such a wonderful person she attracts her friends who are wonderful and make her know how much they appreciate her.

With another therapist in training, who is a year older than I am, he has her with her need for wanting a partner. She was quite obsessed with the idea of wanting the find the right one. As soon as I made this clear to her, she realised she is not in *need* of a partner, she can happily live alone. So if the partner is not the right one, it is not a big deal. She had been struggling with this for over 3 years!!

Another student I have doing **The Course**, who is 42, had no money, no job and no proper place to live. He lived with a guy he loathed and who manipulated him, but he didn't have the strength to leave him. Through The Course and my talking about Ahriman and Lucifer he saw how he had become the instrument of this dark force and so he could change his ways and started to take the reins in his own hands again!!

What you can do now is see what problems obsess you most; where you are losing your energy and then tell him: 'sod you, no chance!' And turn your back on him and focus on your purpose, even if your purpose is 'just' being happy. If you try to be happy without possessiveness, it will most likely work. And it is as valid a purpose as any other purpose.

The Bond of the Dark Forces

I can feel you sighing: what example is there of the bonding of Ahriman and Lucifer? I will give you one example: at the moment very many people want to live more healthy lives. They choose biologically grown products. This is great as it is part of the way to save our planet: but just biologically grown is not enough. If we don't return something to the earth yet only take, it is still not good.

The big supermarket chains who do this kind of biological agriculture but with big machines and enormous amounts of hectares: it looks good as it is bio, yet the whole principle behind the bio is also a fair trade with the agriculture world, the providers, which is not happening here. Plus they don't give anything back to the earth, plus they destroy the structure of the soil with the big machines. That is a bond between Ahriman and Lucifer: it looks good as we take bio, but we don't see any more than that.

Another example: People are aware and realise that bio is necessary, yet the earth is not what it was, so lots of vegetable and fruit don't have the nutrients they used to have, so… they start to take all these pills of greens, and vitamins, for the bones, the liver etc. instead of trying to find the right diet. This is also the bond of Ahriman and Lucifer: Lucifer's work is the making it look like they are doing good for their health by taking these 'wonderful' products and Ahriman is the thinking that they cannot live without them.

Please don't get me wrong, these products are not bad, and when one becomes older it is good to help one's health with some of them, yet a young person up to the age of 50 can normally live without them if she/he eats the right food.

Now what can you do to avoid falling for the dark forces (did you notice I won't give them capital letters?). The best you can do is trying to maintain your centre. As long as you are centred, you will always have some idea, some hold on the right step to take, and it will be far more difficult for him to distract you from your focus.

(I will talk about this in the next chapter: The Heart Chakra.) And of course what I already mentioned before: **reinforcing our openness to inspiration and intuition.** We must learn a new, independent way of judging and think for ourselves, despite the influence of others. Here we will start to increase our willpower to be able to *do* good. For all this you need to discover how to be in your heart chakra, to be centred.

Chapter 7: The Heart Chakra

The previous chapter I finished with telling you to stay centred. It sounds so easy yet how does one do this and why and where, like where is the centre?

The *where* is: in the Heart Chakra. This is the Chakra that we find in our body at the height of the physical heart, yet unlike the physical heart, which we find more at the left side in the chest; it starts in the spine and spirals to the breastbone, which means it sits in the middle of the chest.

The Why? The heart is the centre where our 'Inner Self', or also called 'Higher Self' or 'I-am-Presence' is seated in the body. When we reach this place we come to the unconditional Love, this Love that lives in us yet is not solely ours. We can feel the connection with what some call God or the Divine Forces, The Light, The Higher Intelligence or just That Which Has No Name. Here we are united with the All, so we will not feel alone as long as we maintain our focus here.

The thing is that most people are not in their Heart Chakra and therefore not even really present. This causes them to suffer, whether physically or psychologically which results in the same, yet, when they suffer they are not happy and when they are not happy it is very difficult for them to focus on doing good.

People tend to become more selfish and self-centred when they are unhappy and often fall into the role of the victim. As that causes them to become unfulfilled they often become bitter and want to blame someone for this. They will start to let their rage and frustration out at others, become manipulative and calculated. Now when you think back of what we talked about in chapter two, you can see how devastating this is. It is not the way to grow inside; it is not the way to (co-)create a better world.

When we are in our hearts, we will start to *feel*. We will not be ridden by our emotions any more. In the next pages I will show you some questions from people in a talk, a day after a seminar, and my answers to them. Because I can feel that you have a difficulty already with distinguishing between emotions and feelings, which certainly are not at all the same thing.

You are talking about: when we are in the Heart Chakra. Then we could be unemotional and I find it confusing because of the association between heart and emotions.
Can you say more about that?

I saw that you were very present yesterday in the seminar, and you felt that when you were in the heart chakra it felt actually pretty quiet. Did you feel that?

Yes...

So that is the first quality. The loud noise of the thought patterns fade.

Second quality: when the mind is clear, no thought happening, no attachment happening, a kind of indifference is coming. It doesn't really matter, all those so-called problems you are normally busy with; that which is, is.

That is unemotional. That is simple.

So when people say that the Heart is emotional it is a very wrong description for what the Heart quality is like, but I can tell you why they do that.

When you become emotional the physical heart, the rhythm of the physical heart is definitely changing.

The pulse increases, one can feel one's heart beating stronger than usual, so that's why they say that the Heart is emotional.

But the Heart is totally unemotional. The Heart has feelings. Feelings *are*. (period) It is like Love: Love *is*. Love is a feeling. To differentiate: *being in love* is an emotion. Love causes a quiet state. Being in love increases the heartbeat, quickens the breath and causes this sensation of butterflies in the stomach. We say butterflies in my tummy because the emotions come more from the lower chakras.

With the heart connection with another person, if the other person is not present can the connection still be felt?

Yes, I can tell you what you can do when you feel that the other person is not present. If the other actually *wants* to be present, like yesterday in the seminar we can do the following:
remember when 2 people had to go to their own Heart chakra and then from there make the connection with the other person; when the person is willing to be there but he is not capable, and you notice this, you can actually make the person come down: you go to the Heart chakra and you just tell her/him to come down and you can actually verbalise at the same time what you would like to happen.

However sometimes this is not what you would like to do as the other person might get distracted or insecure by hearing the words. You will be able to decide this, since *you* can see the other person is not in the Heart, therefore *you* can also see whether the person will be capable of following your instructions or possibly might get distracted.

If the other person is *not* willing to go there, you can still make the connection with the Heart of the other person and you can put all the information to the Heart but *the person* can just not connect with *you*, yet you will notice the other person's receptivity or rejection. What you give is noticed in the Being of the person.
You will notice because you will be in your Heart Chakra. If you wouldn't be in your Heart Chakra, if you are doing this just mentally you wouldn't notice the difference, nor would it have an effect.

<u>*After the seminar, being aware of my Heart Chakra, I felt also quite emotional stuff and thought*</u>... (rest retained on request)

You must not forget that when you are not used to being in your Heart Chakra you accumulate quite a lot of emotions in your body. All these emotions go somewhere in your system which might not be the right place and then when you learn to be more in your Heart Chakra suddenly all the emotions start to come up because they are not 'needed' anymore. You start feeling this, and most of what you are feeling are patterns or parts of patterns belonging to the past.

And while you go to the Heart Chakra tears will come because something of this emotional stuff in the pattern is releasing itself and when you cry, you are not emotionally crying; you are neither sad nor suffering: you are just releasing, which you already know as this you experienced today.

You did not feel emotional, you just felt: <WOW!> and that was a feeling, not an emotion. It didn't sweep you away.

What I felt yesterday when I went into the Heart Chakra and connecting with the other people in the room, I felt really good and really healthy and I just wondered; because to do this with all the business of standing alone I didn't feel that I was in any way leaning on anyone.
Do you have any comment about cutting any connection between people?

When you are in the Heart Chakra you can actually connect with IT, the All that Is and 'feeling alone' is not becoming personal. If you personally have an emotional attachment to whatever you will connect with in awareness of the other person, you want to be clear that it is a person you can trust because you are going to feel what is happening in and around the Heart Chakra of the other person.

If you are connecting to the All, you feel the Energy of the All, which is always ok, as it IS, nothing more nothing less. But each person has her own what I call 'expression of their manifestation' and this might be uncomfortable to you unless you can just be in the I-am-presence without the emotional attachment.

When I am noticing that I am more in my Heart Chakra, the more my awareness is going to expand.
I really feel my brain in a way is not active, is not working in the same way as before. It is accessing in a completely different, I don't want to say in a part of my brain but it is accessing something else that I don't really have to think in a way.

In the interview we were talking about the lower and the Higher Mind. The lower mind works through the brain, the Higher Mind through the Heart.
So when you are starting to go to your Heart you can occupy your brain with what you want it to do, which is going with you into your Heart every step of the way and you don't have to be worried about all these thoughts because actually you are accessing your Higher Mind. Working with your Higher Mind means that you might start to communicate differently, without so many words. Like you are doing now, communicating in a way that for some does not make sense yet for those who are in their Heart Chakras it does make sense, as the higher understanding of concepts that are not of this world can be understood in and from the Heart.

Yet here the, what most people call mind, the lower mind, would or might go berserk.

When one is in the Heart Chakra one starts to understand beyond the words. As soon as you start to go to the Higher Mind you can actually talk about things that have no words because with the words we choose, we go around it and try to sort of capture and get closer to what we mean.

There are many things Higher Mind needs translated into words we know, so we will use the library we have in our heads, and as long as the listener is in the Heart, so listening with her/his Higher Mind, she/he can understand what is beyond those words. It works well as long as we stay in the Heart = the Higher Mind. Once we get used to the energy of being in the Heart and working with the Higher Mind, our brain can start to relax, as the brain will normally say 'oh no, I can't do that! - I take a vacation!'

So the brain will work merely to move your intestine or to put your hands up; or telling you that you have to go to the toilet, and will just limit itself to that; which is very pleasant. We will train our brain, our lower mind, to just do that what we want it to do, which is a lot less than what it normally does in a person. Once the brain sees how well it feels this way, it will not oppose any longer.

<u>I realise that I am doing real details in practical things but I am not thinking. I didn't engage my mind. Its quite amazing!</u>

You do all your practical things because you know how to do them. Your mind, your brain-memory, knows where to go, in which department, take the right book, do the right thing. You don't have to be engaged. You can actually be and your body is doing the rest. You will notice at the end of the day that you have a completely *different* kind of tiredness.

It is this 'oohhh, I can sleep really well!' which is very different from 'oohhh I am so exhausted, I don't know how to surmount the day tomorrow!'

I can feel that some of you are starting to experience this also after this seminar. The way most of you feel is your first example: Today was a really full day; you are all tired yet fulfilled and not exhausted as your mind was in peace.

Another example is that when you are in the flow, even finding yourself in stressful situations, you are, maybe against your expectations, not experiencing stress.

Being Who We Really Are

Being who we really are is only possible while staying in the Heart Chakra. As soon as we leave our Heart Chakra, we are acting out our personality. Our personality is the part we notice that is attached to our physical body we carry through this incarnation.
The personality and the physical body are the wrap with/through which our real being, our Inner Self, shows itself to the world. It has taken on a 'shape' through our astrological conditions, our environment in which we were born into and our circumstances, as well as the physiognomy we chose as a body, and all this we have chosen before we incarnated. We chose these characteristics of our lives so as to be able to evolve.

So all of our experiences are self inflicted. We are not victims!!

Unfortunately most people forget this once they get absorbed in and by their daily lives. They only live their circumstances, not the inner knowledge they also carry within them. Yet, by coming 'back' into the Heart Chakra, we start to remember.

I am quite sure that all of you have had the experience in feeling: this, what I am doing/saying now, is NOT me!! Yet out is out, and afterwards one often regrets what one has done/said/caused. By feeling why this happened or where it came from, we often can reach what you then would call 'me' which is the 'I', the real being within you. In the best of cases you will then try to undo the 'harm' you did with your personality and afterwards you will feel good, as the personality feels good after having obeyed the Inner Self (=I am Presence).

The Purpose in/of Your Life.

When we are in the Heart Chakra so many doubts disappear. We can find our purpose in life; this, what we really came for.

For some this might be making music, for another it is being a therapist, for a third person being a good barrister… it does not matter, yet, what you have come for will make you happy while you are doing it. Whatever you do, do it with love, if you cannot, it is not what you must be doing.

When you are not happy at all in your job, you are probably not doing what you came for, or not in the way you must do it.

Feel the passion, as passion points to the direction. Since passion comes from the Heart, like compassion (which is not the same as pity) comes from the Heart, it comes from the place where you *know*. Compassion also comes from the Heart, it is a feeling, pity is an emotion. When one feels compassion, one understands what is going on without the need for interfering. When one pities, one is emotionally involved and attached and… actually in a state of arrogance. When one pities one is patronizing.

Have you ever felt someone pitying you? Do you remember how it made you feel? Did it not make you feel less than that person? It felt unpleasant as the other person makes one feel like one is in a deplorable, low state and they are not, they put themselves above. It feels like one has been judged.

It is so important to find the way to the Heart, as here is the information of which path to walk, which way to go in life. Here we feel the potential we have, which is so much more than what you are living right now!!! Here you can feel that you don't have to behave according to your conditions; you can behave according to who you are which is honest, daring, extravagant, loving, intelligent, straight forward, clean, caring, concerning, joyous and light.

Do realise: your path is unique!!! I can help you *get* there, help you *discover*; *you* will walk it though as *your* path has never been walked before.
If you are following someone's footsteps, you are not walking your path. If you start to believe someone's believes, you are not walking your path. If you constantly recite wise phrases of masters and saints, you are not walking your path.

Your path, as I said before, is unique: you are unique. So think for yourself, which does not mean -don't listen- yet take things in, let them sit and then feel what resonates in and with you. Believing does not help. Observing does.

Take risks; life is a risk in itself. If you only knew how many things could have gone wrong in your life till now; yet here you are, reading this book!! You are alive, safe and sound!

So live!

So many people with fear of taking risks die of anxiety in their bed...
Go, walk, and if you need to, ask someone you trust to every now and then shine a light on your path so it makes it clearer for you where to go. But just a light, don't let them carry you, they might drop you.

Make an effort of finding your Heart Chakra every moment you think of it, every time you remember, even if it is only short, as each little effort will help you to get closer to BE you, and more permanent.

If you are a person that still has an auto-boycott-device 'implanted', help yourself by making a deal, for example: every time you go to the toilet you take a moment to go into your Heart Chakra. Or before you start your car (this you do not very often therefore good for beginners☺) just a brief moment try to reach the feeling of the Heart. After achieving this, every time -I mean even the *attempt*, if reaching the Heart Chakra is still difficult- it amplifies the possibilities, and you can help yourself remembering by, for example, putting a heart on your bathroom mirror and try to go into your Heart while brushing your teeth, shaving or putting make up on. A nice side-effect will be that in shaving you don't cut yourself so often and in putting make–up on you will be more precise☺.

I am sure that you can find many (short) moments in the day to practice this. It is very effective though if you make a habit of starting the day with trying to reach your heart chakra during 5-10 minutes, and before you go to sleep, already lying in bed, you do the same, you will see how you start to feel happier, more settled and peaceful. Once you get the hang of this you will notice that your days are brighter and your nights more restful.

Chapter 8: Enlightenment

So many times I have been asked: 'are you enlightened?' or 'how can I become enlightened?' or 'when do I know when someone is enlightened?' Also some have clearly stated: 'you definitely are enlightened!'

Before I tell you which answers I gave and give to these questions, I would like to ask you a question: why is it so important to have answers to these questions?

Why would you want to be enlightened? What is that fact giving you? Is it going to make you a better person?

If you think that someone is a better person because she/he is enlightened, they probable are not enlightened and you probably don't have a clue of what that is all about.

Is it going to make you happier? Are you more important being enlightened? Is someone enlightened more important than the unenlightened ones? Suppose you are enlightened: then you certainly would not think in these terms. Is it your goal to become enlightened? I must disappoint you then, as you will not reach your goal, when that is your goal.

My explanation is the following: let us suppose enlightenment is one kilometre long, which means 100.000 centimetres. Still supposing: most people are at 1 centimetre... now I happen to be on 11 centimetres... am I enlightened? From their point of view I might be, yet from the real point of view I definitely am not.

Of course I am aware that there is a definition of 'enlightenment' or being 'enlightened'.
For me it is not important to be enlightened, even when some people have defined enlightenment in a way that is not like my own written description. It is not important, to me, to have any title or definition of my state of being.

What matters as far as I am concerned is that each person feels: can I reach a state of Love? Can I maintain my focus in my centre, my Heart Chakra? Do I do all that I do, with Love? Can I see reality for what it is, without having to make it nicer or uglier? Am I willing to live truth, The Truth?

Don't get me wrong, I don't mean to say that you must be able to do all this, I certainly don't want you to get the feeling of having done everything wrong till now, yet your will must be focused on wanting to live this way. This way you will behave and radiate in a way that becomes more loving and automatically will influence your surroundings. You will make a difference; you will be part of creating a better world.

Many people find or make themselves more interesting by setting themselves high goals e.g. to influence the evolution of human beings. Yet too many people set themselves goals, seek these goals, and for that reason become spiritual junkies.

To give you an example: a person meditates every day. At some point, within the presence of a spiritual teacher, reaches the state of being One with All. The person feels this as bliss. This person seeks this experience in all the future meditations, yet, to her/his greatest distress, this state never returns. Instead of benefiting from meditation, this person will become anxious, bitter, depressed, frustrated or something similar.

This does not mean you must stop meditation!! Meditation is the goal in itself so yes, meditate. The interesting thing is, that once we are capable of experiencing this 'togetherness' or the not-being-separate-from-all-that-is, we don't feel it as a state of bliss. We just feel it, that is the reality. It is a very still, very quiet feeling and at the same time wide.

Not wide in a sense of space, as it is not limited by the concept "space" but wide in the sense of limitless in all ways: limitless in space, limitless in time, limitless in the way the mind works, limitless of concepts… just limitless. I would not even say limitless of All, as All implies in a way a limit, a certain totality which feels like having borders. There are no borders unless one starts thinking without staying in the non-separateness. This stillness happens, or is situated, in the Heart Chakra, knowing the Heart Chakra is not a place in space.

We play our part in the human evolution all the time, with all that we do, so if you feel we can or should be more peaceful and loving together, you will have to start to behave accordingly.

Human evolution is something that is happening all the time, so we have to be aware all the time that all that we do has an effect. Lots of the ideas I have given in chapter 2 will help. Living from the Heart Chakra makes this easier and more effective.

In the case of someone setting the goal to become enlightened, that person puts her/himself under a lot of pressure, which causes strain, which drains energy, which causes bad health, which causes (usually) a state of emotional imbalance so that the goal of enlightenment has been driven far away and has become unattainable. Once that person sets her/his goal in wanting to attain calm, the process reverses.

It is all so simple: let go off all those expensive words and just become a loving, generous person. Let go of your believes, all of them. That makes the difference.
From there you can learn to stay more in the Heart and by doing this with full awareness you automatically achieve a state of what is called Higher Consciousness. Don't try to do this the reversed way as that will make the world suffer and you become a prey for the dark forces. Intelligence without Love is a dangerous tool!!

Our goal, if we want one, must be to make our culture a Culture of the Heart!

It is important for you to realise, that living in a loving way, must come from your inner willingness to really be a better person than you are. If you are only acting out because you want to reach this state of consciousness, it is not going to work. The effort/attempt has to be selfless and generous.

You might have a feeling that for you this is unattainable; that you have no possibility of playing a part in this creation of a 'better' world.

That is not true though.

Just the fact that you are trying, the fact that you are using willpower to make an attempt, will already have an effect. Your energy of having the right intention is contagious, and will mobilise others.

You might have been reading books that told you that you must just *be*, and not *try*. In fact this is impossible. If you don't do it by will, you will not really be living what you are.

We must never forget, that, in spite of lots of people saying we are living an illusion, this illusion is a reality in the kind of consciousness we live in. So the best thing you can do is watch this reality. See where you have your difficulties. Watch them without judging and exercise your willpower. Give yourself little disciplines. Not just for the sake of them but useful ones. For example, brush your teeth and use silk thread to clean between them every time, not just in the evening.

Once you get that done easily, take a next one: a glass of warm water as soon as you get up. And a next one: when you have been with lots of people or in the city, come home and change clothes first as to not carry those vibrations through your private life.

You see, you can make little disciplines that are very useful for you and this way you strengthen willpower.

Another way of strengthening your willpower is anything that trains your balance: walking over a fallen tree trunk, walking only on the border of the sidewalk, balancing on rocks, standing on one leg (and lifting your arms, even moving them) walking in the woods and start to move your arms wide open; walk so slowly that you are a long time on one foot etc. You see there are many ways to train this and your willpower will increase.

When your willpower increases you will find it easier to want to, every time, again try to feel yourself in your Heart Chakra. When you do this more often, you start to feel that it bears fruit in your daily life, as you will not react so emotionally nor impulsively; you will feel more calm, you take your decisions more easily, your intuition gets sharper and you will just feel happier. You will have less need to live from your past as you are more and more in the now; and the now is all you have. If you forget to live the now, you live… what? You can only live the moment. Because what is coming is not here yet, so you cannot live it, and what *was*, has passed, you cannot live it.

All these states happen at the same time. Once you reach your Heart Chakra and feel yourself in there, you are living the now. Some call this enlightenment. To me it does not need to have a name, for it separates that what is not enlightened, or those who are not enlightened.

Yet we all are, somehow, on our way to live *what is*. To be quite honest: you can only live what is!!! The only difference is that you notice this or you don't. That is the difference in being conscious or not. Being conscious means you are aware of 'being'. Once you are in your Heart Chakra, you will be aware of Being. And that will be enough for you, without turning you into a lazy person, for once you are here, you will love to share this with others so they can live this way to. And you will feel the compulsion to do good. That is the culture of the Heart!

After having practised your disciplines, you will find that after a while, some of them already are integrated, you won't do without them anymore, especially when you stay in your Heart, as it will become second nature to do what is good for you. Then the effort of doing disciplines is gone.

Chapter 9: Communication

Various Ways of Communication

Writing about communication is not going to be simple, as communication takes place in (almost?) everything we do in our life. We communicate in many different ways, most of these without awareness.

We communicate with our body postures, therefore the expression body language. We communicate with facial features, called facial expression. We communicate with our thoughts; when we do this in an aware state we call this telepathy, yet we communicate with thoughts when we are unaware as well.

Yet the communication we use with most awareness, or maybe I should rather say, which is most obvious to us, is verbal communication, and at the same time this is the kind of communication in which we go wrong most easily. That is why I feel there might not be that much awareness in it after all. We are aware that it is communication, yet we talk without awareness. Here we often, unknowingly, create misunderstandings even though both parties are speaking the same language!!

Good communication can change our world. Good communication requires a whole set of inner attitudes and values, like honesty, integrity, self-knowledge, to just mention a few.

We have to be aware that we are communicating all the time. We have a relationship with something or someone around us, we might be communicating with the plants, the building, the traffic light, God, our neighbour, the washing machine, our partner, our bed, our body, our wardrobe… We are beings in communication.

Now as I said, most of this goes on unconsciously. Why is that?

After having read about the Heart Chakra, you might guess and the guess is right: because we are not centred. We are not in our Heart Chakra.

What happens when you hammer a nail into the wall and hit your thumb? Your first answer now is: I was not centred.

That is right. So I communicated with... Just the nail and the hammer, even with the wall, yet I forgot my fingers. I let the action of hammering happen in an unconscious way. That hurts. You might wonder, why there are relatively few people hitting the hammer on their fingers, after you read this. These people have not decided to walk this path. They are not even aware (yet) they could actually make this decision.

As a matter of fact: wrong or insufficient communication causes pain in some way or another. It might cause you, the doer pain, and/or whoever you are communicating with.

We are so unaware of what we radiate that we underestimate our doings in this world. If we walk with conscious love for nature in the woods, we actually communicate this to all that lives there, and we support the elementary beings in their function of maintaining the woods. In case we visit a place where there is water; a lake, stream, sea, river or well, instead of just enjoying and with that enjoyment only receive, we can literally *give* our enjoyment, our appreciation, to the water. As water has the quality of holding information, this information will cause a lot of good!! It is so simple to communicate well; yet... we have to practice awareness to do it.

At this point I feel some impatience in some of you; you are waiting for what I have to say about communication between people.

The reason why I have started this way will become clear very soon though!

Communication of *Human Beings* with *Human Beings*

When we communicate with a person, no matter what person, we are communicating with a human being. In communicating with a human being we can set ourselves a set of rules, and the first rule should be: **consciously not wanting to hurt**!! This is not the same as not hurting someone, or that the other person might not feel hurt by what you have to say. Yet, if you are careful to communicate in a way in which you consciously don't want to hurt, you are doing it well.

This means that you will not say anything that is not true. Which means phrases like "you are an asshole" and the like, will not be used as no human being is an asshole. You might say: -the way you behave reminds me of-…or is like… nevertheless if you do that because you want the other person to feel bad, you have already disobeyed the first rule. Only say it if you need to state a fact; if not, don't!

Be aware of your facial expression while you are talking; are you having an expression of contempt? If you do, you are hurting the other.

Is your tone of voice full of contempt? You must realise that you have absolutely no right to express contempt, ever!!! Even if someone has done a lot of harm or horrible things to you don't express contempt, as you do not see the whole picture; you don't really know how that person came to do this. So here you have already two reasons why not to show or even experience contempt.

Realise, that, even if you manage not to show contempt on your face, yet inside your personality you are full of it, this will come out to the other person, like a snake in the grass, which means you are hurting them. And you must not hurt others.

Of course you will find yourself in situations, in which you say everything the way it is, you really don't want to hurt, yet what you are saying is an uncomfortable truth, and, if on top of it, it hits an old pattern in the person, that person will feel hurt.

This is not something you can or must prevent. You can tell the person that you are sorry that she/he feels hurt, yet you certainly don't want to hurt them, but the truth needed to be stated. If you are capable of saying things without emotion, you will find that most people, young and old, will accept more easily whatever you are saying to them.
And this is a second rule you can try to practice: if you are emotionally unstable, take your time until the wave is over, and react then. Try to *not* react while you are riding an emotional wave. Even if they are so-called positive emotions like joy or gratitude you might still cause discomfort in the other as they might feel overwhelmed.

Taking Your Time

What is important is for you to realise, is that you don't owe anybody an explanation, nor an answer.

Many of us try to fill the hole created by the expectation of the other and accordingly feel obliged to give a reaction, an answer, instantly. This does not only often lead to misunderstandings; it also causes a lot of what I call verbal diarrhoea. We talk a lot and say little. I know this is not new to you, yet why have you not changed it? Because a lot of this is caused by old patterns you are not aware of.

It is very helpful to first ask yourself: what I am going to say; is that really important? If not, try to let go of it. Yet if you have something important to say and you know what you are saying is true, although uncomfortable, say it.

We also need to learn to say what we want to say and commit to what we are saying. Words like maybe, actually, or speaking in the wrong tense will cause you to talk in an evasive way: I *could* come tomorrow; *maybe* it is nice if we *would* go out; while you really mean to say: I can come tomorrow, and, I want us to go out. Stand for what you say, don't make it weaker, not even because you think you might sound too intense. If you want the other person to know how things are for you, you will have to express yourself accordingly, which means, commit yourself to yourself.

In the times we are living now so much harm has been done in our world towards different so called ethnical groups of people that white people don't even dare to say anything about anyone who is of a different colour than white. I mean than white-white.
In Sweden and the USA they talk about Spanish or Italian people as a different colour although officially they belong to the white race. We don't dare talk about Arabian people, or black people, and are careful in the US to say things like native Americans and who knows what we mean by this? People who never learned history don't.
Are we talking about Indians? In that case why not call them so. Can an Indian person not be proud of being Indian? Just normal pride, like a human being who is glad to be born. The same counts for a black person or a Hindu person etc., as we are all human and all of us have a reason to be.

Yet, don't make such a fuss of it, as no one is better or worse or worth more or less because they are of a certain colour, religion or part of the world.

If we are getting our knickers in a twist by trying to find the right words for what is obvious and yet we try to evade it, we are losing our colour, our colour of being red roses.

If you are genuine, objective and sincere combined with a loving attitude, you will not have a problem. People will be relieved that finally someone calls things by their name.

Being Who You Are

This of course is only possible if you are being who you are; if you are living your potential. And you can live your potential more and more by being in your Heart Chakra.

Being in your Heart Chakra means: being present. Only when you are really present, you will be able to communicate the way communication is meant to be; to cause understanding between beings!

Being who you are also means: living in the present. When I said before: being present, it is logical that that means -living in the present-. How could you possibly be present in the past (☺), or in the future (☺)?

Now you ask: How can we be ourselves?

Is that not an interesting question?

If you are not already conscious in some way, you would *not* have asked this question, as a non-conscious person will not even think about this while a conscious person will try to find out (by her/himself) who she/he is. So if you have asked this question, you are definitely on the right path! This does not mean that if you need help with this, it is wrong. However in the end, only you yourself can find out who you are!

I can be myself by not being a slave of time; by starting to observe. As soon as I stop putting myself under pressure and take time to observe, which means, observing the situation outside myself AND observing myself, inside, my bodily reactions and sensations, I have left time; I am in the Now.

Obviously, when I am in the Now, I am present; being present; being!

When I say observe, I mean the clean observation, which is free of judgments and prejudices. Observing means to watch what *is*. Without speculation. When you observe, and so have taken your time without hurry, without pressure, you will *feel* inside what to do or what not to do.

This is not something you are thinking, this is something that just is, although some people will say: it is something that just comes to you. It feels that it *comes* to you, because the experience of just being is new and it still feels like an action or movement. This causes everything there is inside of you, in your beingness, to feel like new, like something to discover; and therefore again, it feels like an action, although what we could call knowledge, which is what you are perceiving as revelations, is always there and always has been.

From this place of being yourself you will start to communicate with yourself first.

You will not do this as an action, although in the beginning it might feel like this as you are using your mind to get here. You will find yourself in your Heart Chakra, and by just experiencing this, you will find it getting stiller and stiller, and within this stillness you feel Love. Just stay there, live this love, as this is your deep communication with the self and is the only way (as far as I can see) to really learn to love yourself.

When you manage to stay here a little longer, you will feel you are bigger than just what is inside the limits of your body, you will feel you are actually quite huge, extending widely over your body and aura limits. (Here you see that even your aura is not the real you; however it is part of your personality.)

In this expanded state you can feel the presence of what most people will call God, the Divine, That Which Has No Name. You also feel the connection with All that is, whether it is human beings, or the animal, plant or mineral kingdom and all that is far beyond this; this we cannot name as we have no words for it in the consciousness we are living in on a daily base. We might feel it as The Source, there were we come from.

This is the state of truly being what one really is!! This is a state we will experience as peace. The more often we are aware that we can be here, and practice being here, the more we will start to communicate in a way that will cause the world around us to become more harmonious. The elementary beings will start to rejoice as with us, they can fulfil their calling; the plants around you will grow more healthily; the gardens around you will attract more birds and bees; the animals will love you and the people around you will start to see you as different, as someone to notice, and might follow your example.

Of course it is possible that others become more envious or enraged because they cannot manipulate you anymore; this will not matter. If you stay in this 'place' of peace, it will not negatively affect you, and eventually will positively affect them, as you will leave their 'stuff' with *them*, so they only see themselves mirrored.

Of course you can tell them what is happening, if this urge comes from your deep stillness inside. You will, from the Heart, find the right words and the appropriate tone of voice and expression of your face and body, to tell them without getting emotionally involved.

It is important for you not to hold back when you are in Being, when you are acting from your Inner Self, while holding back means you are not being the red rose, you will be acting conditionally as your personality will want you to hold back, behaving like a grey or greyish rose.

When you are being You, your Inner Self, you will not act out emotions. You will feel them, observe them and take them for what they are, yet you will go beyond that and act from your Inner Self, which is compassionate and loving, because you *are* compassion and love. (You might not have discovered this yet ☺)

Facial and Body Expression

We have talked a lot about *talking* now, yet we communicate with all that we are and have. Therefore it is so important to observe oneself.

When we talk, we will stand or sit in a certain way and maybe we have our arms crossed. This way we are already giving the signal: 'I can (for example) make a joke or a criticism but cannot deal with having to receive it from someone else', or 'don't come too close to me'.

When you really want to communicate you have to understand that in communication each communicator must have the capacity of giving and receiving, of seeing and being seen, of talking and listening.

The listening does not just mean listening with the ears, but also with the inner capacities that are called observers, so we observe and take in, we are really being receptive, and accordingly are an active part in the communication.

This also means that if you are really listening to someone talking, you will forget what you wanted to say. When you hold on to what you want to say, you will not listen. Accordingly your face will change, your expression will get a kind of blankness, and the other person will feel she/he has lost you. Therefore she/he will start to try to talk more convincing, persuasive, insisting, intense, in order to get your attention again. The result often is, that the listener closes up altogether, or at least, feels irritated by this way of talking.
Or, if the person who talks is in his/her Heart, he/she will stop talking about the topic and tell you his/her experience of the moment. Unfortunately this last option happens seldom.
Usually, after the intensity, the person will start to feel uncomfortable or irritable and the whole purpose of communication is lost; because the purpose of communication is the sharing of love, which can only happen in a state of willed understanding.

Watch how you look at a person, feel your facial muscles, your body muscles, feel what is tense and relax. Smile, as smiling causes a whole inner state. Do notice: I said smile, not grin!! When one smiles, the mouth is soft, the cheeks are gently lifted and also feel soft, the eyes feel mild. With a grin the lips of the mouth are hard, the eyes get a fiercer look and there is tension in the jaws.

Now check your body: are your shoulders down, nice and loose? Are your hands relaxed; is your sitting or standing posture calm and stable? Especially watch that you are straight.

Having a straight spine gives *you* a better feeling and you look upright, upright also as opposed to sneaky.

While you are observing these details, observe the person you are communicating with; look at the face and what the expression causes in your personality.
Watch her/his body and feel if that causes any discomfort in you, or do you notice that the person is not at ease. If *you* notice that the person is not at ease, be considerate and realise this will influence everything the person does or says. When you manage to be in your Heart Chakra in the Love and Stillness, you will feel whether it is helpful to ask the person whether she/he feels at ease or not. Even if you don't ask this, your state will still comfort the person and put her more at ease.

During the day, again and again watch whether you are relaxed, moving only the muscles you need to and not causing unnecessary tension in your body and face. This state of relaxedness will also improve the life around you that is not human; animals and plants will flourish more and will cause you to be in a state of bliss. *Here* there *is* a state of bliss, as it is the personality that can enjoy this, as the inner you does not need enjoyment; the Inner Self is. (period).

Communication through Thought

We already communicate before we speak or act. In most people there is a mind process, a thinking process, happening previous to their words or actions.

Understand that everything that you think causes a manifestation of some sort. I will give you just a very banal example yet it happens very often, and causes such unwanted results: a young man does not like his neighbour. He thinks his neighbour is a dork. He finds his neighbour stingy, narrow-minded and anti-social. He has never said this to the neighbour, has not even talked about it with anyone. He greets the neighbour in a reasonable normal way, trying not to show these emotions and judgements. Yet, these negative judgments reach the neighbour and in the attitude of the neighbour towards him it becomes visible: the neighbour acts out more and more what the young man has thought about him.

This kind of thinking is equal to witchcraft: we cause our negative thinking to become real, to manifest.

The same counts for positive thinking: have a fabulous idea of how to turn the world into a better place and hold it in your mind: it will somehow manifest.

Here you can see that really all action is communication and what, even in thinking, it can do; what enormous potential, yes, power, it has. As soon as enough people carry a positive thought, it can become manifest.

Hence it always makes a difference what you do and/or think, as it has it's manifestation in the world.
YOU make a difference.

Isn't that a real good reason to rejoice?

Let me give you an example: we take someone who is in therapy. The therapist wants to really help the person but feels a bit overwhelmed. By talking to another therapist, maybe one with more experience, or to the teacher the therapist learned from, all of a sudden there is transformation in the person. Even without further treatment!

This also occurs in family situations. Mothers that feel inadequate in a certain situation with their child, gather, and the talking together in a constructive way causes that they will find their child changed when they come home.

Realise: talking about someone is, in itself, not bad. It depends what purpose you have. If you talk objectively, free of emotion, it can help to dissolve a problem apparently just by itself. Make use of this, as in some situations and institutions it has proven to bear great fruit!!

Chapter 10: Clairvoyance

In this chapter I will talk about different aspects of Clairvoyance and why we need it, and I will also talk about Destiny.

First I am going to talk about: What is Clairvoyance?

Most people know Clairvoyance as the old form of Clairvoyance: the crystal ball, drawing cards, looking at photos, foretelling the future and interpreting Auras, which is the clairvoyance that is still not conscious but a wider developed state of intuition. In this way of clairvoyance the person interprets the signs he/she sees. This is not the way of clairvoyance that I will be talking about.

I will talk about **clairvoyance with consciousness**, and the wonderful thing is: it is something that everyone can learn. The reason for this is that that everyone was born clairvoyant!!

Unfortunately many people usually loose this quality in the first three years of their ever so young lives, and almost all the rest will lose it by the age of 7. But this is no reason for pain or regret, because we can simply relearn it.
I will show you, what you can do to prevent this losing of clairvoyance happening in the children, but...

Let me first introduce you to several different kinds of clairvoyance:

Akashic Records

One type of Clairvoyance is to look at the different times in the life of a human being, the past, present and the future, and this is not just limited to *this* lifetime. This is called the reading of the Akashic Records. This is also the point, where I automatically will talk about destiny, as it is part of what is 'written' in the Akashic Records. In the Akashic Records you will find every possibility that life has for you, or you for your life, with all the choices you can make, all the decisions you might or might not take, and their consequences.

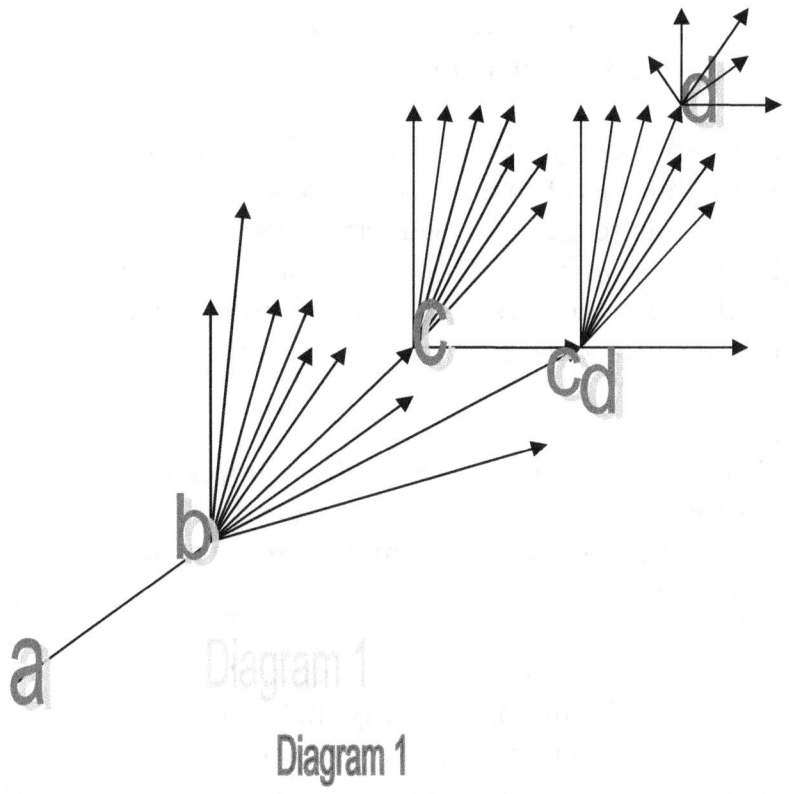

Diagram 1

So if you look at the diagram (1): let us say the moment in our life where we are NOW is at point **a**; we are going forward and come to point **b**: a point where we are confronted with lots of different possibilities; each possibility leads to a point **c**, and from there to a point **d**, yet even **c** and **d** can be the same, depending on which decisions we have taken, we will go one path or another so here it will show how we arrived there.

So we can take the short way, straight ahead all the way, or the long way, with lots of curves and bends… We have the choice, yet the diagram is really quite endless, and therefore shows an infinity of possibilities, each one with its own consequences as we could expand it far more; here I only show a fraction. It shows the abundance of your life, the life that you created. It shows all that you have already lived, with all the choices you made; it shows your present and what you can/might make of your present.

Maybe this already answers another question you might have about destiny: 'is destiny fixed or do we have a choice? '

The answer is yes and no: we have a choice; a fan of choices; yet the consequences of each of those choices are known already; and that counts for each choice we make.

Yet we can always make a different choice, and always change our direction, so the consequences change also.

Now if we really listen to our inner voice, we will find out that we don't really have a choice, yet that we make decisions. The path we must take is clear to us, however often we feel like taking a side road, knowing it is not the best, but just because we feel like it we do so. Here we see that we made this decision. If we are honest to ourselves we don't really have a choice.

You can find all this in the Akashic records, all the choices you might make and all the decisions you can take. All this you wrote in the 'beginning', the beginning of being (human) as we know it, a form of consciousness.

This also shows you once more that it is not legitimate to let yourself take the role of the victim; that is to say that some kind of energy is leading you to do something or go somewhere; we cannot even say that we are a victim of God.

We ourselves made this diagram, from the beginning to the end, we can change our ways, yet we have put down all the possibilities of how we wanted to 'walk' our path through the evolution of humanity; as a co-creator of God.

We have to be aware that the Akashic Records don't only hold the human whereabouts and the human evolution; everything of All can be found there.

As we are, in this reality on planet Earth, the only ones who can take conscious decisions, we are responsible for the WHOLE. This means for the whole planet and beyond.

We can see that if we are capable of reading the Akashic Records, meaning being able to practise this way of being clairvoyant, will help us to walk our path rightly and thus feel more accomplished, even on a normal day to day level. At the same time it will help us to see which way we must go to lead the whole evolution to harmony.

Seeing Colours

The next aspect of clairvoyance that I want to talk about is seeing (a person) in colours. There are various ways to look at someone's colours. If we would look at your aura, and at the moment we are looking at it, you are angry, your aura will show lots of red. This means that if someone took a photograph, for example with Kirlian photography, with which one can see the aura, your aura on this photograph would show red.

That is not *how* nor *who* you are! It is the colour of your emotional body at that particular moment. If at that moment you were feeling happy, there would be lots of green. If you were bored, doped or just waking up, the colour blue would prevail. It is very important to realise that it is only a 'picture' of the moment; it is not who you are.

There is a way of looking at the aura that is far deeper, which shows where and how you are in this period of your life, including the moment we are looking at right now. It shows a certain state of evolution, the colours lie at a deeper layer. I want people to realize there are layers, not just levels. We can literally see different layers of colour.

As well as the aura, which we can see in colour and is the space around us, we have chakras, which are energy points or centres in our body; they are also coloured and we particularly look at the 7 main chakras, the most important energetic centres of our energy system. Their names are: the Root Chakra, the Sacral, the Solar Plexus, the Heart Chakra, the Throat Chakra, the 3rd Eye and the Crown Chakra. In general, when we talk about their colours, we mean the basic colours, but no one's chakras have simply the basic colour. Everyone has different colours, variations on the basic colours, because everyone *has* circumstances in their life (☺) (there is no person without circumstances ☺...), whatever these circumstances may be. And of course we have a past, causing us to have different patterns that will show in colour as well.

Looking at the Chakras we have the possibility to look in 2 different ways. For example the moment that a woman has her period she will have a lot of brown in the Sacral Chakra. This is not really interesting in a chakra reading, because she is going to have her period every month.

It doesn't say anything about her character, personality, behaviour or where she stands in life, so we therefore look deeper; we go further into the person.

In this deeper layer of colours we look to see where the colours are coming from and why they are there, why the imbalance we have noticed, is there. Sometimes this comes from, for example, childhood and we will be able to see this. To heal this imbalance… we can change it with colour. This is the deepest form of therapy, because colour is Light, and we, in our essence, are light. This deepest way of working with colour for healing purposes is called chromo therapy.

Matter, like the body, is a dense manifestation of light. So when we work with colour, which is light, we cause change the health situation in a very deep way.

After reading this I hope you understand that, if you are interested in colour healing, you have to be careful about 'just' using colours, because if you use the wrong colour you can make somebody very ill, as you might be harming an essential colour constellation. Merely taking information from a book and then applying it is very dangerous! Yet if you use the *right* colour, there will be a big positive change in the person, which I feel is of great value. Working with colour in this way, is a very deep way of working in healing. When one wants to do chromo therapy, which means doing healing work with colours, one needs to be clairvoyant in this way first. When we are clairvoyant in these two mentioned ways, we will be able to see what the adding of some colour will cause in the person.

Looking Inside the Body

The third way of being Clairvoyant is being able to see the structure of the person inside the physical body and the functioning of this body. It is almost as if we could see it on an x-ray, yet we see far more and clearer. We can see the bones (including all the joints and the vertebras), the muscles, the nerves and the organs as physical matter *and* how they work and in which state they are.

Furthermore, if there is any imbalance, we are able to see where it came from and what to do about it.

You see that we are getting closer to the reason why it is so important to be clairvoyant.

If we look at medicine: Allopathic Medicine, Homeopathy, Anthroposophical Medicine, Reiki, and all holistic parts of medicine, they have one thing in common: all of them, some better, some worse, are used through the interpretation of the symptoms of the body; Normally practitioners of these techniques do not really know what the person's ailment is. If one wants to be really sure, one needs to find the cause and for this, one has to be objective and able to see. Here it becomes clear, that anyone in the situation of having to diagnose, needs to recuperate his/her clairvoyance.

Not everybody is working in medicine or is a patient; fortunately there are people who are not in need of a doctor. With this I want to say: there are more reasons to become clairvoyant than solely in and for the medical aspects of life.

Practicality on a Day to Day Basis

Now if we come to daily life, and suppose that you are clairvoyant, you will react very differently when you meet someone, than someone who is not clairvoyant. You will, for example, see the happy face that the person you meet displays, yet you can also look behind it, and see what this person is like, or is feeling like, and you will treat this person accordingly.

There are so many people who walk around with a smile on their face, but of all the people that I have seen in my consultation room, the ones that come with a beautiful smile, show that they really try hard to make the best of their life and not bother their surroundings with their torment. I found so much suffering behind their smile, so much sadness, so much distress, so many disasters, so many catastrophes in their daily life.

And I can say: Wow! That is a courageous person, look how brave he or she is, trying to do her/his best to achieve a better state of life. But another person, not being clairvoyant, does not see this, and will not treat these people with the same respect. They might be selfish just wanting to see the happy face, as this way they don't have to deal with the other person's problems.

Because if we don't *see* we might not give the hug that the person needs in this moment. We won't see that the person has a need, maybe it is someone that we really like and like to give to but we can't do it because we don't know/see. If we are able to see, we will want to give the person what she/he really needs.

Of course this also works the other way around: if you meet someone who is clairvoyant you will receive more of what you need than when the person is not… and also in a way you might not like (☺). Here I mean to say that you might get told something about your attitude that does not go down so well as you get told that that attitude is not truthfully you.

Here comes even another aspect of clairvoyance: Life is not just full of friends.
We all work in this life, or at least have to deal with people in their work environment, and everybody in some way or other manifests their patterns, at different times and in different situations; with which I mean to say that people show a side of themselves while they are manifesting some part of their character or personality, that is not necessarily pleasant to others. And because there is a society that we have created, but don't feel very good in, we meet people that we don't feel very good with.

There is a lot of mistrust, a lot of envy and greed, to speak of a few qualities and it will be very useful to know whether the people we meet are the right people for us, or are people that we cannot trust.
We do have our intuition; we *will* have this little feeling; yet we cannot really see, and the danger that our mind interprets and misjudges is high. How often do we have a first feeling (intuition) and then discard what we felt as what we see *appears* to be different from this first feeling…

Clairvoyance in point of fact is not intuitive but knowing, so we will know that what we see is not the real thing, and our first feeling will get respected as it is, meaning it is true and not coming from a whim.

This doesn't mean that we will not have problems when someone is going to betray us, because even when we see this is going to happen, sometimes this is inevitable, yet we will at least know beforehand. This causes a different inner reaction.

It happens sometimes that you want to trust someone whom it would be better not to trust; if we are truly on a spiritual path, we very much like to trust people and to see the best in people. Giving trust invites others to trust us. Living our mistrust means receiving mistrust.

If we 'only' have intuition, we will often override this first thing we see (the reason to not trust). Yet if we are clairvoyant, we see and remember (not meaning to become resentful!!), because we have it so much clearer than this vague presentiment of intuition.

Do you already feel that clairvoyance is not just something for mystics but a very practical quality for your common daily life?

The last part of clairvoyance that I will mention is about how it all started, what is God, what am I, how did all of this creation come about?... I will talk about this in a later chapter as it deserves a chapter of its own.

Chapter 11: Clairvoyance and Children

Losing Clairvoyance

Now let us go back to when we were very young.

After the child is born; at, let us say, 18 months old, he will learn to walk and, as a consequence of this learning process, often falls. This is a normal process; he is learning to walk.

One day he comes home crying and sobs to his mommy 'I fell and my knee really, really hurts!' and mommy says: 'come darling, I will give you a sweet and then a big kiss on the knee and it is all over; the pain is gone.' Now we are going back to the feelings of this child... For him his Mum is like God; she knows everything, she is his whole world.

She is usually the most important person in the life of the young child, (together with the father and often the grandparents). But Mum said -it doesn't hurt- yet *he* feels a tremendous pain in his knee... so the child is thinking... 'Ok Mommy said that it is all over and mommy knows everything so... I must be wrong'.

The child starts to learn not to trust its own perception. That is where the child starts to lose its clairvoyance, this is possibly the first moment of the great loss. The child grows up and is going out to play in the garden -he is still Clairvoyant-; he is looking not only at the plants and animals but also at some beings, beings in the plants, the trees, in the water. When he has contact with one of these beings, it often happens that the child feels like a friend, and so the child will give this little being a name and these elementary beings (this is what they are called) love to be in the company of children and love it to be given a name by them.

And so the child called this particular being Leo. The child comes back into the house and says to Mommy: 'I am going to play in the garden with Leo, who is the little one living on the tree' and Mommy says: 'There is nobody living on the tree!' 'Oh yes! Leo! I play with him every day!'

Now mommy does not want to upset her child and thinks she needs to play this game too so she pretends that she understands, tells the child it is fine and then Mommy says: 'I have a little friend too, she is called Agatha, and is right here with me cleaning the couch'... The child says: 'Mommy that is not true, there is nobody there!' and looks with this frown on his face of no-understanding.

Mum insists. For the Mum it is a game, she does not realise what she is doing. She thinks she is doing a good job to go down to a level that she thinks is the level of her child, but for the child it is not; for the child she is behaving like an idiot: he really *sees*!

This is not a very helpful behaviour of the mother's part! When we are in these circumstances it is better to just admit we don't see anything and *ask* the child about his friend; maybe even ask him if we have such a friend also, although we can't see nor hear him or her. But the result in the above mentioned example is that the child starts to lose trust in his own way of perception, he loses trust in his mother and he will start to lose his clairvoyance.

Let us look at this from the beginning. The first stage of incarnation is conception. Then the second stage of incarnation comes: a child is born. When a child is born, he/she is clairvoyant. Always.

At the age of two and a half or three the third stage of incarnation occurs, which becomes very noticeable as the child starts to say 'I' instead of calling himself by his name.

At the age of seven, the time when the child changes from milk teeth to permanent teeth, the next part of incarnation takes place. This also is the moment that the mental capacity of the child starts: the child can think. Before this age the process of thinking did not really take place. The child was an emotional being, without mind-movement.

In case you don't believe this, try to tell a child below this age anything you want or don't want it to do: you will have to repeat it again and again, and if you understand how a child 'works' inside, you will try to make it clear by a way of physical action in the child, so that the body starts to memorize it, if not, you will have bad luck, as the child will not get it.

This is the tragic in many families as the parents don't understand their children's process and think they are just being disobedient. Children under 7 don't think or reason, although sometimes it looks as if they do/can; they cannot make a thinking pattern until they are around seven; the age they are starting to lose their baby teeth. Before that they are organic and emotional.

Thinking

Now let us look into our grown-up world. Those who try to walk the spiritual path notice that the most difficult thing is to get the mind to stop. Whatever we try, meditation, reading, learning, talking to someone; most often the mind is in the way, and what is most disconcerting is that the thoughts repeat themselves.

It is difficult to stop thinking. Someone wants to give you good advice and says: 'don't think so much!' Yet how do you do this?

Watch the daily life from a distance: most people watch quite a bit of television, spend a lot of time behind a computer, children spend time at the computer or Gameboy. All this means: a constant input of images and impulses and even just going into the city is a constant avalanche of impressions, which is a lot of input for the mind. Therefore a lot of our capacities remain in the background, as we splurge a lot of energy on the mental input.

Let us go back to the child.

In our educational system, the child often gets forced to start learning to read and write at the age of 3. From what I have written before, you know that this is impossible for most children, because the child is mentally not ready for that. This means that the child, under the pressure of the system, will have to find a way to use its energy for that. It needs energy to grow yet now it needs a lot of energy for something it is not ready for, so it needs to let go of something that might get in the way of what it is learning: the capacity of *seeing*, truly seeing! Because soon enough it notices that this is an unwanted quality: the adults accuse the child of lying, of deceitfulness or fantasising. This is very sad: the child looses the most important quality it has besides love.

(Also physical health might, even unnoticed, deteriorate)

By the time the child is seven, it has learned to shut out everything in the way of the mental process. Consequently it starts to think; finding logical patterns, without any joy, because the real thing has even left the background; it disappeared altogether. The child has become a 'worthy' member of our society. How awful.

Now there is no reason to totally despair.

The good thing is: you can relearn it!!! Everybody can relearn it.

In the trainings I give, I offer many ways to regain the clairvoyance you have lost and even develop it a lot further.

When you try to play an instrument, you will discover, that, to be good at it, you need to practice as much as you can. In becoming clairvoyant this is the same: the more you practice, the better you get at it. If you stop practicing playing your instrument, you will not be a good musician any more. The same counts for clairvoyance, if you keep on practicing, you will be able to see; practicing means you will not lose it.

The practices I offer are not difficult; and you don't need extra time for them. The important thing is that you should be practicing all day long, so I have developed disciplines you can integrate into your normal daily life.

Doing this, and by gradually practicing different disciplines, you will first become more intuitive, and gradually reach clairvoyance.

If you are interested in these trainings, you can get in touch with me or simply look at my website in what I have written about THE COURSE.

I would like to share with you some of the questions people have asked in my conference about clairvoyance: one of the questions was about children.

A woman asked <u>how she could ensure that a child will not lose their clairvoyant abilities.</u>

The first thing we have to realise is that we must always be truthful, totally honest with any child; all the time! When you are not feeling well, and you don't want your child to worry about you, then say: 'I don't feel well, yet you need not worry.' Never say: 'I am ok', when you are not. The child sees that you are not, yet if you say you are, the child will either not trust its own perception or not trust you. It is of utmost importance to be truthful.

When you are truthful, your child will learn it can trust you. The child will feel: mummy tells me her truth, and this is the same as what I see. So the child feels safe and understood.

If you notice your child looks thoughtful, ask: 'did you understand what I just said? Do you see the same as I do?

Do you think there is something different from what I said?' It is important to keep on checking because sometimes it happens that the child sees something different from what you see, therefore it is important to always ask.

This is also important when one has a partner and one makes love. Almost always at some point the child is aware of your lovemaking, might peek in and see things it does not understand. When you openly talk about what is going on, making it clear that these are things that grown-ups do as an expression of love, yet that for children this is harmful, the child understands, because it sees the pleasure and the danger, yet cannot place this paradox without you mentioning it.

For a child it is also vital that, when you are angry with a person you love, you tell the child exactly that: that you are angry with the person yet have not lost your love for her/him, so the child knows, ok, this is something that can happen; another paradox. As a side effect: for you it is good training to actually realise you *do* love the person you are angry with, as, having this in your consciousness, the outcome will be different.

For a child it is important to see that many situations are ambivalent and that having ambivalent feelings or emotions is ok.

I have often been asked *what kind of education does a child need*?

I recommend Waldorf schools. The teachers of these schools have been trained in the understanding for the phases of development of children. They don't start mental education before children are 7 years old and they put a lot of emphasis on things that induce or maintain clairvoyance. They work with the elementary beings even if many of them can't see them; they work with colours, learning and memorizing through physical movement instead of sitting still, music, and a lot with art. It is a world where nature and art are intrinsically connected and lived. Of course it is possible that individual teachers are not to your liking, like everywhere else in the world. Yet it is a stately recognized system and the best I know.

Often people ask me to *quickly mention something easy they can start doing right now to become clairvoyant.*

My answer always is: be totally honest, especially with yourself. Look at yourself with objective realistic eyes. By doing this you are on the path.

Another question: *If you are older is it more difficult?*

No, it isn't. I have people in The Course from 20 to 82 years old. And the 82 year old man is now definitely clairvoyant.

Another advice I can/must give you is: if you smoke, stop smoking. If you want to be clairvoyant you will have to stop smoking, because smoking in truth puts you in a cloud; the cloud of the visible smoke, yet also an invisible cloud which lays itself on your mind. You are hiding, yet as a consequence of having this cloud around you, you cannot see either. This counts as much for normal tobacco as for grass, hashish etc. When one smokes one tends to only see very close to the self and therefore looses the overview of what one's own attitude causes in others. It is hardly possible to see the consequences of one's behaviour when one is enveloped by a mist. So if we cannot see the 'normal', the obvious behavioural consequences, how will we see the deeper layers?

(The next question came from a mother who has a son who is clairvoyant, yet uses it as a weapon against his mother, knowing she does not see.)

What can I do if I have an adolescent who says he is clairvoyant and therefore he is always right?

If none of the parents are clairvoyant it is a very difficult situation. Because as soon as you mistrust him when he is truthful, you harm him; yet if you trust him when he manipulates you, you also harm him. If you feel that what he says is truthful just trust, yet, if you think he is manipulating you, just say: 'I think you are manipulating and I don't want that'.

Try to increase your intuition as much as possible, and become clairvoyant yourself. Children are smart and they will come to an age when they will want and search confrontation to learn to live in this world, it is like practising real life. This is not an easy period for parents, and many parents of adolescents become too permissive at this age, as they are afraid of losing the love of their children; these children know this very well. It is not productive to live this fear. Here you will have to set your limits clearly: 'this far and no further'. Be firm, even strongly severe. It is good for the child, and for your relationship with the child, as it gives security. What you have to learn is to love your child *more* than your fear of losing him.

Severity, showing a child boundaries, actually creates security. A child has to learn from these boundaries; that everybody has their own space and the right to live in that space. Realise that in this way you are helping your child to become a good person. If you just let the child overrule you, the child will become a very unpleasant person.
Unpleasant people meet unpleasant situations, and therefore will not attain happiness. I am sure, that you want to witness your children becoming and being happy.

Question about if reading at the age of 4 is permissible.

Some children have a great capacity of understanding the written word at a very young age. Of course we will not tell the child off for this. If a child finds joy in reading at the age of 4, we will let it read. Yet we will not support it in a way that we will help the child to read more or better. When we do things with the child, there will be other things besides reading.

Without pressurising, we will try to occupy the child with creative manual activities or music.

In my own case, at the age of three I learned to read. As so many times I had been read to from one particular book, I knew it by heart; every time I looked at the book, I started to associate the separate sounds with the separate group of letters I saw in the book. So from the group of letters, the spaces in between, and the repetition of groups of letters coinciding with the sounds in the same 'rhythm', I learned to differentiate words. In my case it was, at first, unnoticed, so it was not suppressed nor supported. Therefore I also spent lots of time doing other things. The consequence though is that I still love reading. Yet I also love doing manual things, especially if they are creative.

Question: What is happening when you become clairvoyant?

The way I teach it you will learn all levels of clairvoyance at the same time but you can choose what you need. This does not mean that you are constantly shifting your awareness, the shift happens automatically. Just remember when you look at a landscape view over a lake, at the other side are villages and mountains. When you want to see that particular little castle, you shift your sight automatically so that the castle becomes more focused, while the rest of the view moves into the background.

In the beginning intuition will start to increase. You feel something rather than actually seeing it, this is how it begins. What you will notice is an increased sensitivity, for example whether you like being or not being in a place, even your place at the table. You will clearly know whether you want to be with a certain person or not, which road to take, where to go shopping, what to say in a particular moment etc. You may find that by changing a plan and taking a different road from usual, that on the usual road an accident happened.

You can check these things. Also things might happen like you might feel that a red car will come around the corner with a woman at the steering wheel. Then look: probably this is so. Do realise: clairvoyance becomes first noticed by feeling things, yet the difference with just feeling is the certainty knowing 'they are'.

<u>The next question was: what is the difference between instinct and intuition and intuition and clairvoyance?</u>

Animals have instincts, it is like the 6th sense, and animals need it as it is their survival tool. By destroying certain patterns in nature, like the magnetic field, changing the causes of the rivers and streams, building big antennas etc., we destroy part of the instinct. So instinct is not a permanent quality, it is dependent on certain conditions and it can be destroyed.

Intuition is a state that cannot be destroyed. It is an input from a higher level of consciousness. It is being connected with All. Intuition can be trained, and even if the conditions are unfavourable, intuition will remain, as long as the person remains calm. Intuition goes beyond the survival sense, it has a deeper, more spiritual and nuanced quality to it. Intuition is a feeling: what is right, what is wrong. Instinct is not a feeling; instinct is a sense causing instantaneous reaction. Intuition does not necessarily cause a visible reaction, as we decide what to do with what we intuit.

Clairvoyance is one step further again. Clairvoyance is knowing. Clairvoyance is intuition made conscious; it is the logical consequence of developed intuition.

The intuitive person might have many negative voices inside, which are still very loud and affect the person negatively and cause discomfort. But someone who has developed clairvoyance has an inner security, that if these voices are present, she/he will check to find out what they are, and/or if they are real.

I will give you an example that happened many years ago. Someone gave a talk in a hotel in Ibiza, Spain and the conference room was full. The man in question pretended to be born in Orion, the star constellation. I was thinking 'This is really rubbish' yet I noticed to my agony that the words that he used were great.
He had the audience hanging on his lips, his talking seemed to make so much sense, and the way that he played with the audience was truly fascinating. I started to worry and thought: 'Oh my God! What is going to happen here! If people believe this man, it is terrible!' (in those days I still worried about that).

105

At one point a couple, who were about 80 years old and were sitting behind me, stood up and said as from one mouth 'Everything you say sounds nice, but it is rubbish and we are leaving.' This shook the audience awake and many people started to applaud. The old people used what we call Intuition. The rest of the people were caught up in what I call the mind game: the game of rhetoric and semantics.

If you have good intuition, the step towards clairvoyance is not large, yet you must have the courage and will power to give up all your beliefs. You have to start by only wanting reality, even when this reality does not appeal to you.

If you manage this, you are strongly on the way. This also means that when you are meditating, you must not use your imagination: if you are meditating on a chair in your living room, don't pretend to lie on a meadow with wild fragrant flowers, with a stream running through it. This will stop you from becoming clairvoyant and will only enhance you seeing things in a distorted way and not the way that they really are.

Realise that the beginning of becoming clairvoyant, for many people, means, that they first start *feeling* things, not actually *seeing* them. You can also become clairaudient, which I put in the same category, as it is the same way of evolving and developing your innate qualities.

One last warning: you might not even notice how much more awareness=clairvoyance you have acquired, because this process is so gradual that you get used to it very quickly, each step of the way. So you might not see the difference as soon you take it for granted feeling or seeing this way. This might sometimes discourage you, because it does not feel spectacular. Yet most real changes happen in this way: little by little... therefore durable. And... on the whole this is quite spectacular!

Chapter 11: Meditation

Meditation and Centring

There is a tremendous need for centring. Most people feel this as a need to meditate and hope to get, as a result, inner peace.

Unfortunately most people, when they start meditating, learn some old technique, and often this same technique drives them up the wall instead of calming them down. They have been taught to close their eyes and to try to see a blank wall. Yet all they see is hundreds of images and, through the strain of trying to see this blank wall, they start to think more than normal. So much for trying to achieve calm…

Others are taught to see darkness, and when they achieve this more or less, they panic as this darkness frightens them.

It is difficult for most people to concentrate on nothing… not knowing what nothing is. How is that with you? Are you able to still your thoughts? Most of you, when you are honest will have to admit you cannot. This is not a crime, nor a drama, just very human.

What is Meditation?

Each Meditation teacher will give a different definition of the word Meditation. Whether a definition of the word is important or not, does not, to me, really matter. Yet, I would like to explain to you what I see as meditation and why it is important, how to go about it and where not to put more importance than it deserves.

One of the first things I call meditation is the being aware…

What does this mean? Whatever it is you are doing, as soon as you are doing it consciously, with awareness and you are aware of this consciously, you are in a meditative state. This is very important; it means you are making meditation a way of life, or, life a way of meditation.

This means that if I were to give a definition for the word meditation, I would say: meditation is being consciously aware of being conscious. I know, you will have to say this a few times before it makes sense, yet… you will see, Meditation is not just being conscious, as *that* is not enough.

I could finish the chapter here, yet I am aware that you would not be helped by this alone. I will give you some advice that will help you to get to this state of meditatively living.

Helpful Ways to Start

When you are waiting in a queue, instead of trying to evade the looks of other people and let your mind wonder, start to very consciously observe the way they are dressed, the colours of their shoes, the wristwatch, the clips in the hair, the greying at the temples, the dyed hair grown, showing the real colour at the roots, and so on, yet do this without letting you be drawn into judging it; don't think whether you like it or not. Just watch and take it in. This way you have more capacity of perception, as your emotions are not getting in the way.

Later in the day, for example before you go to sleep, try to remember this picture fully, with each and every detail, and if you know there is something you don't remember, yet clearly feel the lack of something, then complete the picture with something you feel fits in the place of lack. This is a very good discipline to be more aware of your environment and it sharpens the memory and the capacity for concentration, which is a help even in driving your car or doing book keeping☺.

Another exercise: when you are somewhere outside, try to hear all there is to hear. Normally we have a selective hearing, and we choose, subconsciously, to *not* hear many sounds. Close your eyes if you are in a place where being with closed eyes is not inconvenient, it makes it easier. Yet if this is not possible, do it with open eyes, and, instead of trying to let the sounds fall to the background, try to hear them all. Later on write them down and see how many you have remembered/noticed. You can make a game of this doing it with others and compare what you heard; maybe you realise when seeing lists of others that you did hear something they heard, yet were not really aware of it. This will sharpen your senses and will train you to not want a different reality from the one that is…

When you do the dishes or wash your car, check all the time which muscles you are using and which muscles you would not have to use and try to relax these.

When you drive your car, stay with where you are driving and don't think of what you will do once you reach your destiny. This you can apply to all sorts of other situations also, the same as the previous one. So I mean to say that when you are occupied with something, simply pay attention to what you are actually doing and don't waver or drift with your thoughts. It will help you also to become more focused, to finish your chores faster, to be more efficient, to have less problems with fitting your program into a crammed time schedule etc.

Another helpful practise: Sit still and take one word: for example eye: think all you can about this eye: it can see, it is a ball, it has several layers, lenses, it can feel burning, it can have cataracts, one can close it with eyelids, there is the inner eye and the outer eye, it might need contact lenses or glasses, one talks about the evil eye- the third eye- a black eye etc. One can read with ones eye. One can have different colours of eyes. Usually one has two eyes... Go on until there is nothing left you can think of concerning 'eye'. Every time you catch yourself thinking of something that has not a direct connection with 'eye', come back to 'eye'. I call this: conscious thinking within willpower. It helps you to organise thoughts and train and discipline the mind.

Once you can do this with easy words, use words that actually help you with your inner growth. You can use words like discipline, willpower, unconditional love, selflessness, responsibility, generosity and also the negative side to become more aware of the traps: envy, greed, abuse, (sexual abuse, abuse of power, abuse of position) self rightness, etc.

This way, while you are disciplining the mind, you are evolving, becoming more aware of your patterns, yet, while doing this, stick to the exercise and don't let yourself be tempted to go into your new discovery, as that is getting emotionally involved. Trust, that once you understand something, you really understand it. Of course the sitting still in silence is an important way of starting meditation, but you might want to start with these practises first when you find it difficult to have stillness inside.

Once you feel more at peace in your mind, start with 15 minutes sitting still with your eyes closed, focusing on your heart chakra and don't move your body. Over time, when it gets easier, you can increase the time you do this. But if you don't have time to sit still for an hour, do the previous exercises, they will and can help as much.

Visualisation

A common practice in meditation is visualisation. Whether this is good or not depends on what you mean by this. When you mean to bring something into view, to start picturing something that is happening in your meditation, it is fine. Yet if visualisation means to imagine something that is not there, it is not right.

It is very important to NEVER imagine anything. Stay with the reality. If you are sitting on a chair in a room meditating, don't imagine you are lying in a field with wild flowers and that you hear the sound of a stream running close by. This is Lucifer's work trying to take you away from reality. Always stay in reality. If you don't see, don't imagine. If you are receiving a guided meditation just feel if what is said resonates somewhere inside you. Feeling is also a way of seeing. If the guide use imagination techniques, do something else, go inside and watch your body from inside.

If you have a still, non-guided meditation, find a focus. A very good focus is your heart chakra. Yet if this is difficult, make it your physical body. Start to do a body check: feel the feet, first bones, muscles, blood and lymph vessels, nerves, tissue, skin, hair, nails etc. going up through the body this way, feeling the organs until the whole body is done. Then try to stay with the feeling of relaxation that this has caused and, to not fall asleep, sit straight up, always spine straight, without leaning. This way you will stay awake or wake up again when you tend to fall asleep.

Yet, even if trying to get into your Heart Chakra is difficult, keep on intending to do this. In chapter 7 I have made it clear why this is so important. Here you come into consciousness from the place of love, where you prevent becoming an intelligence without 'a heart'.

Even the attempt to get into your heart chakra, again and again, will make you become more who you really are, what you meant yourself to be when you created yourself. Now I know that this is not so easy. Therefore I will give a few more ways to make this step become easier and easier, once your mind is more disciplined, so it will cooperate with your inner being instead of wanting to do its own thing, and wanting attention while doing it.

You can start feeling your physical spine, and, if you have any knowledge of the 7 main chakras, just feel the part in the spine where each chakra starts and spirals to the front, or, in case of the root chakra, to the bottom. It is a good idea to start with the root chakra working your way up or just choose a different chakra for each meditation. Feel the third eye chakra between the eyebrows and the crown chakra on top of the head. First feel them physically, and maybe, while staying with one of them, or one after another, depending on your endurance, you may find yourself coming more towards the inner, non-physical quality of the chakra. Once you have tried this, gaining more experience, you might find it easier to reach and stay in your Heart Chakra.

To come to this place where you can perceive consciousness without making it personal you can practice the following forms of meditation:

You can feel your thoughts and just watch them without thinking them further, and after a while try to feel where the thoughts come from. Just keep on exploring: where do thoughts come from…

You can observe your thoughts and feel the space between two thoughts and stay in this space; after some practice maybe you are capable of making this space between thoughts larger.

Whatever kind of meditation you practice, it is important to just begin, as once you meditate, you will find more calm, more peace in your life. And this already has a positive reverberation on your surroundings. Be careful not to set a goal, only the purpose of wanting to become more aware, more loving.

You see, there are many ways of meditation. Yet the most important really is to be aware of how you interact with people, doing it from your Heart, from the place of love.

Meditation is important, yet, if you make the sitting down away from your environment a way of life and forget to consciously take steps to become the best person you can possibly be, you have missed the point. So many people become Vipassana junkies, running to each retreat and yet never really advancing but prance with the fact they are really spiritual as they do this practice.

One is not a better person because one does some sort of discipline. Yet, meditation can bring calm into the body and soul that causes a person to become more self-controlled and more considerate. It can cause inner space in which revelations can well up. It helps to still the seemingly endless flow of thought-patterns that keep us captive in certain ways of behaviour. It helps one to become less emotionally attached.

Here I am talking about the meditations in which one withdraws from the normal busy circumstances of daily life and retreats in a quiet place for let's say half an hour or an hour, to seek stillness.

If one does everything consciously, and one is conscious of being conscious in every act, the whole day long, while living a practical life in which service is a protagonist, then yes please, make meditation your way of life. This way meditation is an integral part and not a form of escapism.

Do try to find those moments of stillness, as it will make it easier to *not* put yourself or others under pressure nor will you feel inclined to talk too much. You will develop your inner senses as well as more compassion. Your sight for real beauty will deepen and your understanding for real values will increase or return. You will feel happier, your anxiety will decrease, and you will be a better companion to yourself and others.

Maybe you have noticed, maybe you have not, that, while you are reading this book, there is something happening in you although you sometimes might not understand what I try to convey. There is a deeper understanding happening and this is happening in the Heart Chakra. You will notice that, when you start practicing meditation in whatever what way, you will perceive this deeper understanding in a more conscious way.

I will give you an example from my own experience: I work hard and have many projects going on.

From the outside I am certainly behaving like a good person. Yet from inside myself I feel this stress, this pressure and uneasiness. I have looked at this many times and made my life quieter, more moments to tank some energy, and to avoid pressure of all kinds as far as this was possible. Yet this feeling persisted and I could not give it a name. I looked at the feeling of not being good enough; the feeling of missing something I should do or say or remember as this was how I felt. Yet this did not make a difference.

I meditate every morning and every evening. All of a sudden in a meditation I felt that this unease is guilt: a feeling of living too good a life; I have a lovely husband, wonderful friends, enough clothes, shoes etc., lovely daughters, I do the work I love, I get acknowledgement and enough money to buy the food I really want to eat; I can choose what to eat…

I have been in Africa where it is heart wrenching to see that so many people are not only in the position of no choice, they cannot even eat the amount they would like to eat, have no clear and clean water to drink. They hang the laundry over plants, not because that is their custom, but because they don't even have enough money to buy a laundry line, or pegs.

That is why I feel guilty; because I live a good life. I have been trying to maximise everything that has to do with work or effort, so as to not enjoy this wonderful life. I would live very spartanic, eat more simply than I like to or that is good for me, not read a good book, but limit myself to reading what enriches me for work; to sleep and rest less than I *really* need, and not going to the sea as that is free time… do you see what I mean? What a waste??? Having such a wonderful life and not allowing myself to enjoy it? Yet this is a pattern that comes from the fact that when I was born my father did not want to let me live, so my first blueprint was that I don't deserve life; let alone a good life.

Meditation brought the bits and pieces of the puzzle of this part of my life together over the years, by going into my Heart Chakra I was/am in the middle of what I am, and started more and more to see the different parts of my life, of my patterns and circumstances, from the inside; from the part that is the real I. If meditation is done correctly, with the right attitude within and not as a hunt for pleasant sensations, it is of very practical use indeed.

When you have chosen what is best for you, I recommend that you meditate twice a day, every day: first thing after waking up and last thing before going to sleep. Times that are particularly inspiring and strong for meditation are dawn and sunset. Realise that meditating before going to sleep might induce sleep and your meditation might not be longer than 5 minutes. If you have difficulties falling asleep, this is helpful. Yet be careful not to become oblivious: meditation is about being conscious!!

Also realize, as soon as you are making it a constant to try to be the best loving person you can be, you *are* in a meditative state!!

Chapter 13: Horses?!

You will be very surprised to find that I want to write about horses. Maybe you feel you are not a horse person at all. That might be true because you have not had the opportunity to meet horses the way we meet them. That is why I will share with you part of what living with horses means to me, as I have seen that we can learn so much from them, if we observe them closely.

The horses I will talk about are Icelandic horses. The reason for this is that these horses are the ones that are usually still kept in a natural way and they are possibly the purest race on earth, closest to the original horse. As I feel that some might object to this exclamation I will explain it briefly. As the law in Iceland has decided long ago that no horses might be imported to the island, and every horse that gets exported must remain away from Iceland, they have kept the race pure.

Icelandic horses are not put into separate stables; they live together in herds as they would in the wild. This creates the natural behaviour that I want to talk to you about.

Every herd has a leading horse. Some herds exist only of young stallions, some of just mares; some herds are mixed having male and female horses together yet usually without stallions. The reason for this is that stallions will battle with every other male horse to possess the mares. If the herd is mixed, there will be a leading male and a leading female horse, and, depending on how strong the character is, one of these will even be more dominant than the other. Yet, normally the mare is the leading horse and will be followed by the rest of the heard. She has the wisdom. She takes care of the food, protection against intruders but also protection of the weaker ones within the herd. She will educate and also will make it clear she wants her peace... peace above everything else.

Take the situation where new horses need to enter an existing herd. Suppose they are mares going into a mixed herd. One of the new horses has been a leading mare all her life, is a strong mare in passing on her characteristics in the breed and is very strong in character. A horse with a strong character means that the horse is in balance inside, at rest with itself, has wisdom, and other horses will almost always acknowledge this wisdom.

This last quality of the horses in itself would be something we humans must aspire to... Acknowledging wisdom in others without envying them. Life would be so much more beautiful and efficient if we could and would not let our egos roar...

Back to the horses: they are let into the field where the existing herd is. Let's say it is a mare and her female foal. The existing herd will be curious and come and see. As this horse has been a leading mare she will walk away, her foal following its mother. This will cause the others to not attack. After a couple of days the new mare, which has seen that she is wiser and more capable of leading the heard than the leading mare in function, will approach the herd. They most likely will refuse her and put their ears toward the back, which is a hostile attitude. She will withdraw and over the next few days she will gently show her capacities, coming back towards the herd yet never challenging the leading mare, as this would put the herd in danger. (She cannot know that real danger does not exist not being in the true wild.) After a few days one or two horses will make friendship with her. From now on they will seek her council. More and more horses will come to her side. Once this is clear the leading mare will come and make friendship, yet probably will make it clear to the rest of the herd that she is still, after the new leader, the highest in rank. All this gets done without a fight, without hurting, yet with clear wisdom to show ones capacity and not holding back and starting to sulk.

Horses make it very clear that there is a natural hierarchy and that only when the natural hierarchy is acknowledged will life work well.
This is something that we humans still need to learn.
Why do we want to lead when we are not born leaders, why envy a person who is?
This does not mean that the herd will stay the same all the time. The hierarchy can definitely change as horses grow up and become stronger, more experienced and wiser. The disputes amongst horses who are not the leaders are frequent, as when horses grow up, they change... like humans, and become stronger or not. These fights for the rank are normally harmless as they, the horses, find their protection and well-being more important than winning. Once they see they are not the strongest, they will let go without losing their pride. What is, is...

When just one new horse is introduced into the herd usually all horses want to reject it, which they do by making the new one run. Yet there is always one horse, not necessarily the leading horse, which will run in between the new horse and the rest, to protect the new horse from bites and other attacks. This usually leads to normal herd behaviour within an hour, sometimes within minutes. It is fascinating to see that the other horses respect the decision of this one horse to protect the new one, and that they don't go for an attack, which they easily could.

Another interesting way of reacting to behaviour is when a horse behaves wrongly, for example when it is too aggressive or too dominant at the time of feeding or, when it behaves badly towards the weaker horses. First it will be put in its place by a horse that it has challenged or in the last two cases (too dominant at feeding time or unfair behaviour towards inferiors) by the leader, which can be done verbally, with squeaks and whinnies, or with body language: ears towards the back and sometimes the hint of heaving the back legs (just the hint of kicking out) or in the case of males also the front legs. If this does not work the horse might get kicked. Yet what happens mostly is that the leading horse will turn its back on the horse and gather the rest of the herd, leaving this horse out. That is a clear message: if you cannot behave well, you cannot be part of our 'family'. Only if this horse will apologize, which means it will come hesitatingly with its head low and chewing, it will be allowed back into the herd.

What we can learn here is that fighting is not a way. It is better to just make clear what is not good and ignore one who does not want to respect this and only when our fellow men behave well, can we let them (back) into our life.

Yet what is perhaps the most impressive experience, when one works with horses, is the fact that they are soo tremendously strong: in their neck the smallest horses have more than 7 times more muscle power than any strong adult male human, yet, they are so noble. They will not use that power to hurt, they would rather flee. When I see what some (unfortunately far too many) people do to make a horse docile so they can ride it, which is the breaking of horses, I often think of what could happen if the horses were like humans… with the power it has in its body we have absolutely no chance. Yet, it prefers to have its will broken than hurting us. Is that not the lesson Jesus the Christ wanted to convey to us?

Now we must not think that when this has happened to a horse it will be our friend. We will have a horse that obeys, but we will never have a horse that will think for itself and therefore in a situation of real danger take over and save our life. It will do as we say yet we will always feel that it does it because it feels forced, not out of sheer pleasure of being with the rider, with pleasure of running or walking. And it has learned to always mistrust humans.

However, if we work with the horse in the horse's language, it takes pleasure in what it does, it will want to please us and be our friend. And when we come to this place, that we try to work with and for the horse, not against it, we will be able to trust this wonderful animal and it will trust us, and we will be able to go wherever we want, as the horse will let us know what it likes and what not and where it feels danger and where it enjoys the run.

This way we have a companion that is equal to a friend, just the language is different. Yet it is important to let the horse know who is superior. In human relationships this is no different. Instead of fighting for acknowledgment or recognition we must just stand firm in ourselves and be what we are. Then our place is clearly defined and usually not challenged. Of course, as with the horses, with humans there is also, again and again, the challenge of wanting to see who is 'stronger/better'; but as the horses do, we must also do: just stay being what we are and not fight, not take the challenge as an invitation for battle: one cannot kick-in an open door!!

Of course I could talk about other animals too, as they all have their particular qualities we can learn from. Yet to me this animal, always noble, stands out, as it will not by itself turn itself against us, unlike most other animals.

Discrimination

When I look at human beings, even at myself, I see how much we discriminate. Not just race and colour, also the way people dress, have piercings or tattoos etc. When I looked and observed horses I did not find this. I have seen mixed herds, meaning all horses being of different kinds of breed: Haflingers, Icelandic horses, Arabs, Hannoveraners, mixed blood, cold and warm bloods all together, and there was total harmony.

Or herds where there is one race like the Icelandic horse yet all different colours and… again no problem.

Now don't think horses don't see this difference between them, because they do. I often observed that horses of the same colour become real great friends, or a very big horse with a tiny Shetland pony; or horses of the same breeding line, yet bred in very different parts of the world, who will seek each other out as friends although they have never seen each other before... Sometimes there is disharmony, yet not because of different breed or different colour; it has to do with the behaviour.

What about us humans. We discriminate all the time and often will not admit this. Yet it is very important to be aware that we do. If I walk on the streets being a white, and I am meeting a black person and try to show my best side because I don't want this person to feel discriminated, I am categorically discriminating!! It does not matter when this happens if it happens in good faith. We don't want to discriminate; we do feel that all humans are equal and should have equal rights, yet if we feel uncomfortable with someone of a different colour, religion or culture, or even if we don't notice it is uncomfortable, yet we clearly react because of the fact she/he is different from ourselves, and we act accordingly, we have to realise we *are* discriminating, in spite of ourselves. As long as we are aware of this, it makes it easier to react properly and put our fears (as this always comes from fears, eventually) to the background or aside. When we open ourselves up to the person in this way, we might get beautifully surprised!!

We must realise that all this is happening inside ourselves because of a program we carry with us subconsciously. We don't have to blame ourselves for this as it has its good reason, yet we must not be superior out of discomfort either, as we are not. If necessary or if we feel we trust the situation enough, we can state to the other person what is going on inside us, as often this helps to relax the situation, and… the other person might be able and willing to help us to overcome this uneasiness.

If you would ask me to tell you what *I* see we can learn from horses altogether, I would put at least the following qualities we should learn from them if we really want to create a better world, as only a better world for *all*, is a better world at all:

Force without control; firmness without hardness; clarity without coldness; tension without becoming cramped; attitude without rigidity; softness without weakness; beauty without vanity; adjustment without resignation; magnetism/charisma without manipulation.

When I read these last few sentences, I kind of go WOW, if we would only get to a few of them, what a different world we would have attained already!! Just imagine: all the strong characters of this world not trying to control, what an enormous power would be available for the good, for the whole all of a sudden. If you yourself are a forceful person, try to feel how less strenuous it is the moment you let go of the need for control!

Firmness without hardness: this I feel especially thinking of any form of education. We teach with firmness and are consistent, yet stay loving and never getting hard, how wonderful this is for our pupils, students, children... they would find security and feel supported, never having the feeling something or someone is *against* them. It will make them want to learn eagerly!!

Clarity without coldness: when we use our -capacity of seeing- to facilitate, being constructive, coming from our Heart Chakra, from the place of love, we become of great help and are a great force for the new society we try to help create.

Tension without becoming cramped: this is being like a cat before it jumps: all alert yet totally relaxed but not limp: the whole organism is awake. We will be ready for action at any moment, nevertheless when action is not required we will be relaxed and not waiting, yet totally engaged in the moment.

Attitude without rigidity: we will be proud within what we are and will show this, yet there is nowhere even a hint of wanting to defend what we believe, live, render and have as an attitude. When new clarity comes, new insights, the attitude transforms because of them. This way we will bring our values into the world so it can have a look at them, and accept the ones it finds valid, and we do the same with the new values we get exposed to.

Softness without weakness: when we are soft, we are pleasant for others. They love to have us near so if we have something to convey they will be receptive as our softness gets experienced as something that leaves them free and at the same time embraces them. When we are not weak within the softness, we will radiate a strength without oppression, we will not be manipulated nor show or live any kind of victim role. Our presence will be very convincing without us trying to convince.

Beauty without vanity: when you look in the mirror in the morning and you see your beautiful face and body, you can have this feeling of awe; 'this is me and I am so grateful for this! That I am allowed to look like this and move this way...' You will rejoice in your looks without presuming anything. In what you do: when you are doing something and you see how well done it is, how perfect, beautiful, special you have done it and just rejoice in the capacity of creating, seeing and living such beauty, without having the feeling of superiority, no chip on your shoulder, no conceit, you are of great benefit for this world, even when this world is not used to this yet; it will learn!!

Adjustment without resignation: we are living in a society and are not alone on this planet. There are things we will do to have the community function well, to not create disharmony, so we will make adjustments. Yet this does not mean we give up ourselves or what we believe in; it means we will mould into the community to make it function to the best of its possibilities with our entire will and approval, giving all we can give and holding back there where this is required (still).

Magnetism/charisma without manipulation: some of us have this special radiation that attracts. The art is to recognize this and employ it for the good. We will attract people and with totally open cards, we will let them know our intentions. We will let them be totally free to believe and decide what they want and, even if we are convinced it would be the best for the whole, will not use our qualities to persuade them to anything at all.

What a task we have before us, what a wonderful, interesting, adventurous task. And I am not even talking about the equality of gender yet: as I can see that if we have reached equality of gender and accept open sexuality (I don't mean lovemaking in the street!) so that the taboo is off, first then we will be able to really change the world into this paradise we all so strongly long for. So women, don't boycott your own gender: listen to your 'fellow' women and don't stop listening when a man starts talking (even interrupts). This of course counts for men too: listen to a woman when it is her turn to speak and don't allow men to interrupt or behave as if they are priority of have preference in existence.

Accept female bosses/managers with the same naturalness as male ones. When you are a woman: Stay a woman with female qualities and don't copy men's attitudes for that is not staying true to yourself and the reason you chose to be a woman. This is only the beginning of a large yet vital process...

Look into your own mind to find out how much you are, through inherited behaviour, accepting the roles of male and female behaviour. Do you in your Heart also find it normal that men often talk in a patronizing way to and about women? And this question I direct to men and women!! But also be clear: where there are qualities in which men or women are simply better, let them do this!! Search in your head and your Heart and see how much there is that is or is not in unison. This in itself is a grand way to achieve greater consciousness.

Chapter 14: GOD AND I

We have so many questions and ideas about God. Yet I have been very brazen and named this chapter God and I. why would I do this? What is there behind this? Is this really just being brazen or impertinent or does this have a deeper meaning?

I hope that after having read all the previous chapters you trust there is a deeper meaning in this.

Where It All Began...

When I want to know what God is, or what I am, I have to go back a long, long way. Now unfortunately most people have lost the capacity to go back inside of their memory that exists beyond this lifetime.

As I worked hard to get this back, and during many many years I have been looking back in what we call time, I will tell you what I have seen. I very much realise that this might be untrue for you, that you will not want to believe this, which is just fine. Yet I ask you to please read the whole chapter to see if there is something that resonates within you, or even triggers something within. I also want to make clear, that my first time of looking into this was many years ago, yet every now and then I do it again, and find no change in what I see, except for maybe more details, which for this chapter are not relevant.

Most of you reading this book will know that most people believe 'it' all started with the Big Bang. Yet these are suppositions. I have tried to find one person that actually has seen this, experienced this, and found none.

I can only trust what I experience, and what I see. So I decided to take on the challenge of going back further and further, until I could go no more. I am going to explain what I saw and experienced, share it with you and hope you will not laugh at me. I will use words we know to try to describe my experience and sometimes this is hard, as for most of what I went through there are no words. I will try to convey behind the words, what really is...

The whole time I experienced *my* presence; it was different from when I am clearly in my body doing things, yet I can state I was all the time conscious(ness).

I experienced something I can best describe as fog; some kind of dampness without feeling it as water, being light and darkish at the same time; a bit like being in a fog and someone shining a light on it from afar. Yet this felt more like as if it were darkness holding the capacity of light in it, yet it was not really dark either.

There was something I would call Joy of Being, or maybe Joy of Experiencing Itself. And Love; all had this presence of Love, sweetness and blessedness in It. Yet, at the same time it was very still; still as the absence of noise. And I was there, not as that which or who I am now, yet I was 'it'. You were 'it' too. All I know and intuit to know today, was 'it'; was 'there'.

It was very still yet carried a feeling of constant bubbling, excitement, in it; and something I would describe as curiosity. There was a sense of desire for expression but not in the way we feel, not with compulsion. It was totally free of emotion. Yet there was something like a desire for showing 'itself' its own possibilities. The 'urge' for creation started to become manifest.

Now please realise that we feel emotion with almost every word; the word urge, the word stillness, the word joy, we connect them with an emotion. Yet here there was and is no emotion. They are feelings, and feelings are. Without any kind of judgement to them.

So it created wisdom. Now wisdom was given a name: Sofia. (This is the Greek word for wisdom). Within the giving it a name, a word, IT decided to first be idea, then word and then manifestation.

Now don't get me wrong: the creation and 'it' (is also you and me) is all the same and yet all of a sudden into a forward motion all going on simultaneously. (it is soooo difficult to describe this which is 'real consciousness without mind' with our vocabulary!!) This all was happening very still, very soft and very strong at the same time. When you remember of what I talked about in the previous chapter: softness without weakness, this definitely was that: Softness without Weakness. And I, the I-am-presence, was there!

I am going to make big jumps now, as I don't want to go into detail. Sofia in her wisdom first 'mentioned' light, and light manifested. Then Sofia created what we now call the solar system yet it did not look like what it looks now. Only planet earth already has undergone many changes since then. I won't even mention the others. And we, in or as the I-am-presence, were there all along. We were not physical, yet already were destined to be humans or better said: the human evolution, and we were already part of a hierarchy.

We were starting to become more diverse: hitherto I cannot call it persons but something similar to that, as expressing it otherwise becomes too difficult to understand. We were a manifestation of light in different forms.

And even in this state of being we already had dissonance. Not particularly with or towards each other, but we found within ourselves characteristics we needed to develop. We would now call it weaknesses, however in that state of our evolution that kind of emotional values were not experienced yet.

To remind us we needed to develop these qualities, we created the animals, each kind representing a quality-in-perfection we must reach at some point in our evolution. Here I could see that 'we' already knew that one 'day' we would be like we are now, in a body, needing to respect the animal world and learn from it.

I am not going to explain the whole experience to you, for that might look like I want to convince you of something. That is not the point. What is important is that when we started to become a *we* instead of an *I*, we did not need to talk, we were not physical. We would 'think' to each other, could commune with each other by becoming one again, like dissolving two different colours one in the other, and we would know…

And… I, the I-am-presence of each of us, was here from the beginning; creating all this we have now, all along the way. O yeah, you read this well: you were there!!! Right there at the very beginning you have created, like you are creating right now. There was never even the slightest doubt you could become a victim… you were there and decided!!

You are here now and decide!! You are the creator of your world, you are God!!!

This is blunt, brazen??!! Ok, yet it is true. You and I and all of us are God. Altogether we are one and are creating. If we hurt someone, we DO hurt ourselves, as we are causing a worse situation in this world, which means a little less nice world for us, ourselves…

As soon as we understand this well, we can use this tremendous power we are for the good: we can create a real wonderful paradisiacal world.

I am It, It is I, It is We, We are One.

I can make this chapter into a whole book, yet I have the feeling that if you read this a few times, you will understand and see it. Even if you find the first part a bit strange.

Do you deny we are all one? Can you see we are one with All? All is the whole planet and all it includes and… all there is beyond.

So if we are all One, all each of us does, reverberates everywhere. We are responsible. On top of it we are privileged as we can decide the direction evolution takes. We are responsible, also responsible for that direction. We are the only species that can decide direction. We wanted it to be this way; we created it this way. Now we are not allowed to pretend we don't have that responsibility as we ARE the *only* species that decide. Not taking responsibility means we will destroy our own creation. We will destroy ourselves, always, for we are ONE.

It is wonderful to see how powerful we are and what all we can do. It is so easy to change the world if we, humans are willing.

Just think of God as the sun. The sun is this enormous ball of light and its rays shine on all and everything and make it grow (in all kind of ways). Now each of us is a ray of this light. Light is there to shine, not to hide. It would be ridiculous to create light for it *not* to shine. So we *must* shine. That is our purpose. Now do you remember the beginning of this book where I talked about the red and the grey roses? You have to be the red rose that you are, because that is shining the light. If you behave like a grey rose, for whatever what reason, it is not valid, for you are hiding your light and light is there to shine.

When all of us shine our light, all is beautifully bright and no darkness can exist.

Is there more to say? Maybe I can say that we *are* not the sun, but the light of the sun is in us and we are in the light of the sun. You are not God, but God is in you and you are in God.

Who am I? What am I? Is there a Difference?

Does this answer the question 'What or who is God?' for you? Or the question 'who am I?'

What I would like to suggest to you, even if you have understood everything in this chapter and even totally agree, to ask yourself again and again: '***who*** am I?'

Go deep, as deep as you can, because you *are* not a sculptor, or a doctor, or a teacher. That is NOT who you *are*!! It is what you perform.

To give you an example: the first time I asked myself this seriously after having seen the above described evolution of the world, I saw I *am* Love and Compassion. Not a teacher, doctor or therapist.

Then ask (like I did) '*what* am I?' please don't think that the *what* and the *who* can be the same: they unquestionably are NOT!!

Again, go deep. Listen quietly inside till the answer rises up, manifest, reveals itself. The answer I got was: The Manifestation of Love and Compassion on Earth.

These same questions I asked myself every day twice and during a whole week I got the same answers. Yet the second week the answer to the *who was Love, Compassion, Understanding and Patience,* and the answer to the question of *what* am I was: *the imperfect tool to transmit these qualities to humanity* ☺.

After more than another week the answer to 'who I am' was 'Peace and Tranquility', and the answer to the 'what I am' was 'The Intention of Manifestation of Peace and Tranquility for the Human Form'. And each time I felt I got a bit further in achieving the manifestation of my own divine creation.

When you read this chapter, after having read the previous chapter, does it all make sense to you? Can you see how important it is to learn from the animal world; to integrate or better said, to reintegrate the animals into our world?

I really hope so, as I find it tough to fill pages and pages with the same thing, explaining the same thing in different words, pretending you don't understand. I don't want to be one of these authors that fill a whole book with one sentence expressed in many different words. I cannot do that. I just want to write a book to help you with ways to get there. I don't want to pretend you are stupid or even ignorant, because if you were, you would not have come to this chapter. I quite trust you understand, as I also trust that you know that when you reread parts of this book, what you read will reach a deeper dimension within you; will touch different parts of your subconscious that want to become conscious. I totally trust your sincere endeavouring for what I call your divine self. So I keep things brief, and you can use the same material over and over again, as you will start to discover more and more meaning in these words, you will acquire more and more understanding of you, of the 'I' and of 'It'. We will shine our light together and we will feel no separation, no rivalry, no competition. We will be happy to shine as we are united in the light.

How I See God and Myself

When I concentrate on seeing me, what I really am, my essence, my I am presence, I first see myself as something like a liquid, kind of transparent, 'substance' that is shaped like a filled dome, or maybe a ball, and through it, somewhere in the middle, I see my physical body. The body is not very big, maybe 5 percent of the 'space' I take up. This liquid substance comes from and goes to the middle of the heart chakra. It looks a bit like when one pours liquid and it makes this funnel shape. Yet at the same time the 'emptiness' of the funnel is also full with it. Yet the body is in it, and at the same time part of it; penetrated through and through by it.

When I keep on looking I see that I am so big that this 'liquid me' actually goes all over the world, over this planet. It is like a layer of some 3,5 - 4 meters.

Then I want to see what God is. I hear it/her/him as a voice within this liquid-like substance. This voice is all through this liquid-something that is the essence of me. The interesting thing is that I actually *see* this voice, not just hear it. And it is not just hearing and seeing either; there is this presence of voice all the time. The attention-grabbing thing is that it is experiencing God within me; yet at the same time it is me within God, it sooo feels like one.

Yet there is clearly this voice that tells me what to do, what not to do. I would like to say it gives suggestions, yet it is not really like that. It feels very different from what I am and at the same time, if I decide not to listen, it definitely feels like being unfaithful to my own being. It really feels like the great consciousness is my own consciousness at the same time. It makes me very happy to be this Godness as much as being part of it and I feel this responsibility of our creation also in my hands. It is very interesting in how wonderful responsibility can feel. When I am right here, very aware of what I am and not distracted by trivialities, responsibility is nowhere a burden; Just a fact that we all have and need to take care of.

I stay still in this experience and then it almost feels as if I were the front person and I would have the whole world, especially the energy of humanity behind me in a kind of a membrane that I pull forward with my energy of evolvement. This way any move I make, inside and outside, moves this Whole behind me with it. It is not something that I actually *do*, it is something that happens. Yet… it is a choice I have made to do long ago, long before I incarnated into this body I wear as Ria.

This is the chance, the responsibility, the opportunity we have by being conscious.

When you watch a tree, you will see that at some point in its years' cycle it will shed thousands of seeds. Yet just a very few of these eventually become trees. There is a lot of growing power put into the perpetuation of the species.

When we have an idea to help humanity, and we carry this actively inside our hearts, when we then act upon this with willpower, there is a lot of willpower put into it, which is not being 'used up'. This willpower does not get lost; it is not wasted energy. Whatever is not being used here on earth goes into the ether, where it is spend on humanity as a whole.

In other words, your willpower goes into the bank for the benefit of all humanity.

Here you can see how just a few of us, if acting correctly, can really change the world!! We don't have to become missionaries; with just a solid group of people we can actually achieve a whole change-over. This is being and living our true potential. Let us all make use of it!

Chapter 15: Epilogue

Live as if you are celebrating

Every negative thought hides a fear. It is because of fear we stay trapped and we procrastinate and avoid making necessary changes; it is because of fear that we don't take decisions; it is because of fear that we allow things to happen that we don't like and we don't stand up for ourselves.. Fear makes us feel powerless; it separates us from our true being; it enslaves us, it diverts us from being happy and keeps us from experiencing peace and fullness.

Every day is a new day that we create. We created the possibilities of living and dying. We created the possibility of experiencing joy and misery. We created the possibility of choice. We created a sensitivity and subjectivity and so we can experience all previously mentioned conditions. And... we can create these conditions at random.

Now if you are still in the victim role, you will not agree in this, yet I will tell you then that you created your victim role; it is your choice to live that now, so if you want to perpetuate this situation, you will experience misery and every new day might be a prolongation of your torture.

I have decided that every new day is another opportunity to feel the beauty of creation. Every new day I experience my possibilities and investigate which of them are due to be shared with the world and which ones need cultivation before I can let them out. I also check which parts I feel as myself, yet are preventing me from being happy and what I can do with them, or let them be, in order to change my life so that my light shines more effectively.

Does this all sounds like a chore??

Not if you take on the attitude of grace. Be grateful for your life, it is a celebration of creation. Every part of this life, of every day, of every moment, of every experience: look at it as unique, for it IS unique. When you look at it from a distance, with objectivity, you will see how interesting the smallest thing that you do is.

Watch yourself cutting with a knife. Look at all the movement that is required to do this, and all the body functions that have built up to make this possible. If you do this you can only rejoice in this magnificence and then you are celebrating.

See the sky, how many varieties of sky have you seen in the last 5 minutes... let alone in your *life*. This also counts for those living in places where the sky seems to be always one colour, as it is not. It moves, one can see the movement in it when one really looks. Teach yourself to see, to feel, but now *really* teach yourself.

Not to perceive pain because it interferes with your calm is a choice you can make, but if you really want to be happy you'd better choose what you want to see and feel, and stay with it. It does not mean you have to stop living your 'normal' life, but you will find that your life is not so normal after all; that it is very exciting, as the smallest, apparently insignificant movement or occurrence, is preceded by this enormous magnificence we call creation. And... creation is I. You, the I am presence you are, is this creation. The You is part of all this, so celebrate. This is not narcissism, this is true love.

My life is a happy life, in spite of having had quite a lot of severe moments in my life; some of these in which death already took me and I decided to come back. These experiences also restrict my bodily movements. However, I am alive; I can see, move, eat and experience my inner growth, becoming a better person every day. I can see that my endeavour to evolve internally bears fruit which causes me to feel satisfied that I am on the right track.

I see people changing for the better and becoming happier under my guidance. This all gives me reason to feel that living this life is celebration and it causes happiness, as I know that, even if it might take quite a while, still, in the future, this oneness will be more imminent and then all people will want to follow this path of light, the path of love.

We are on our way!!!

Words of Gratitude

When I think of whom to thank, so much gratitude towards Rich, yeah, you Richard Scanlon, comes bubbling up, as you have supported me and helped me with your correction in such a loving and respectful way that I could not think of anyone who could have given me more support and could have done this, possibly very ungratifying job, better than you did. You helped me to believe that all is possible. I can write my book, yet am terrified of anything to do with paperwork, new programmes on the computer, and all the bureaucracy that trying to publish a book brings along. Yet there you are and were, all the time inspiring me, explaining things so it does not become a burden. Thank you!!!! And it is not even finished!!

I am also grateful to The Powers, all these huge and small beings that are difficult to name as not many see them, but have been with me all the time and kept me going or reminded me of doing the, hopefully, right thing. All these powers together we call God, which let me be One and therefore made it possible for me to know where people have difficulties and start to experience more separation than Oneness so I would know how to write this book.

I am grateful to Pete Bampton as he was the one that said: 'you need a book in which you share your wisdom and experience' and therefore I started this book and left my autobiography aside. Thanks to Erinda and Sharon, who are honest with me all the time and wrote me their first feedback. And last but not least thank you my dearest husband, that you love me and support me unconditionally and that you cooked for me when I was absorbed in writing.

www.ingramcontent.com/pod-product-compliance
Lightning Source LLC
Chambersburg PA
CBHW081047170526
45158CB00006B/1884